FREEING
ENERGY

FREEING ENERGY

How Innovators Are Using Local-scale **SOLAR** and **BATTERIES** to Disrupt the Global Energy Industry from the Outside In

BILL NUSSEY

For information about this title or to order other books and/or electronic media, contact the publisher:

Mountain Ambler Publishing

ISBNs:
978-1-7325446-3-5 (hardcover)
978-1-7325446-4-2 (softcover)
978-1-7325446-5-9 (e-book)
978-1-7325446-6-6 (audiobook)

Printed in the United States of America

Interior design: 1106 Design
Cover design: The Book Designers

*This book is dedicated to the thousands of innovators
who are in the clean energy industry today and the
thousands more that will soon be joining.*

*May you help free energy from a century of fossil fuels,
outdated monopoly business models, and technology stagnation,
and make cheaper, cleaner, more reliable, and more
equitable energy available to billions of people.*

TABLE OF CONTENTS

CHAPTER 1

IN SEARCH OF ENERGY FREEDOM

NINE DARK MONTHS

In September 2017, Hurricane Maria, a Category 4 juggernaut, hit Puerto Rico head on, weakening only slightly as it spent eight grueling hours ripping the island apart with 100-mph winds and heavy rainfall. Sixty-four people died during the initial storm, but it was the aftermath that proved most deadly. Puerto Rico's entire electric grid was out for weeks, and many areas, especially those in the lowest income rural communities, had no power for almost nine months. It was the largest blackout in US history.

In the end, an estimated 3,000 people lost their lives due to disease, lack of medical care, and heat exhaustion, among other causes, laying bare our society's utter dependence on electricity.

Power outages happen all the time, so what made this one so tragic? It was the extraordinary duration of the post-Maria blackout. A few hours without power are an annoyance we are all familiar with.

A day or two might require some adjustments. But weeks without power can be punishing and months can be deadly.

Like many places around the world, much of Puerto Rico's water system depends on electricity to power pumps. Without it, faucets ran dry, toilets would not flush, bathing became impractical, and clothes were left unwashed. Lack of refrigeration caused medications and food to spoil. Medical equipment went dark. One woman told me that her brother had been in a hospital on life support when Maria hit. The hospital's generators worked for a few days and then failed. Her brother passed away before they could be repaired.

Across the island, people were cut off from each other. Roads were washed out. Emergency supplies could not reach victims. The most basic services ceased. Gasoline stations could not pump gas. Eighty-five percent of cell phone towers remained off-line for weeks, largely due to lack of power.[1] Elevators stopped working, cutting off upper floors from anyone with physical limitations. Disease spread; emergency services became overloaded. Puerto Rico became an island of generators, meaning already economically devastated communities were forced to pay for diesel fuel, one of the most expensive ways to generate electricity.

Two years after Hurricane Maria, I traveled to one of the hardest hit places on the island to hear personal stories from the people that endured the multi-month blackout.

The story of a school and its community

Perched high up in the central mountainous region of Puerto Rico, the town of Naranjito rises about 3,000 feet above the Atlantic Ocean. As I exited the highway from San Juan and began to climb the steep roads into town, the dense urban landscape gave way to the tropical forest of Puerto Rico's interior. Some nearby fields had been cleared, where many of the town's 30,000 residents farm crops like coffee, tobacco, bananas, and, of course, the oranges ("naranjas") that gave the town its name. My destination was Escuela Segunda Unidad Pedro Fernandez, a K-8 school nestled in the picturesque mountains of Naranjito.

Célines Pacheco López, an English teacher at the school, described what she saw when Maria's winds finally died down: "My uncle's house right next door was made of wood, and it was completely destroyed. They lost everything." She explained, "We couldn't get out of our community because we were blocked off by a mudslide, so we had to dig. Everybody with a shovel started digging. When we got out, we found that another mudslide had taken out the main road, and it had collapsed. It was three to four days after the hurricane before we could finally get out."

Even after the ordeal of digging out her home, she was still the first teacher to reach the school. "When I got here, I couldn't open any of the gates because all the trees were blocking them. When we finally got the gates open, it was a disaster inside." I asked her how they cleaned up the school. "We're an all-female staff. We were ten teachers dragging a limb across the street, dragging pieces of metal and garbage, everything."

Without electricity, the school was forced to shut down. This was due to the lack of sanitation more than any other reason—the school is built on a steep slope and electric pumps are required for the water to flush the toilets.

Like so many Puerto Ricans, the 300 children who attended Pedro Fernandez had their worlds turned upside down. Many were traumatized, having lost their homes. A few lost loved ones. Others watched as friends and family members moved away, hoping to find a better future on the mainland. Parents had to quit their jobs to stay at home and look after their children. One teacher told me, "It was six months before all the electricity was restored. We basically lost an entire year of education. We are doing what we can to make up for the lost time, but it is very hard."

A few townspeople had generators, but these were never designed for continuous use. They were also very expensive to operate. With the town's roads destroyed and the island's ports overloaded, the price of scarce resources like gasoline and diesel soared, compounding the financial duress of the hurricane's survivors.

But I did not come to Naranjito just to chronicle the problems its people had bravely faced. I was also there to witness a solution. I wanted to see firsthand the school's new breakthrough electricity system, which promised to protect the school and its community from future outages, whether they were due to monstrous storms or the island's notoriously unreliable grid. This system made the school an energy pioneer and it offered proof to Puerto Rico and the rest of the world that there was now a better, more resilient way to power our homes and buildings.

Pioneering a new era of electric resilience

As we walked through the campus, children laughed and talked all around us, oblivious to the amazing technologies that were silently powering their school. Looking down from the mountainside on to the roof of a nearby building, I could see the shiny, deep blue solar panels that were capturing Puerto Rico's ample Caribbean sunlight. But what was truly remarkable about this system sat below the panels, inside the building. Mounted in a closet were a set of nondescript boxes containing batteries and electronics. Together, the panels and batteries create a system called a *microgrid*—a tiny, self-contained version of the bigger electric grid that powered the island. Now Pedro Fernandez, like a few other schools on the island, was prepared. The next time a storm hit, the school would be ready.

FIGURE 1.1 *The Segunda Unidad Pedro Fernandez K-8 school located in Naranjito, Puerto Rico. Credit: Bill Nussey.*

When the Pedro Fernandez microgrid was first brought to life, it was a surprisingly unceremonious affair. The lights did not even blink. But its impact was profound. With the pull of a switch, the school and its community were suddenly independent—freed from the shortcomings of the island's grids. Lights, computers, kitchen appliances, and other basic necessities could be powered locally with nothing but the sunlight hitting the roof.

If disaster struck again, the batteries and the solar panels, which are designed to withstand hurricane force winds, would ensure the electricity never stopped flowing. And, as the school would learn a few months after my visit, the microgrid would continue operating even after a 6.4 magnitude earthquake shook the island.

The origins of the Pedro Fernandez microgrid project are as inspiring as its impact. Led by RMI, formerly called Rocky Mountain Institute, several local organizations joined forces in early 2018 to design, fund, and build the school's microgrid. Roy Torbert, one of RMI's leaders on all island-related projects across the Caribbean, explained why microgrids are important for building community resilience in Puerto Rico's rural schools:

> The hurricane was particularly disruptive to the lives of the island's school children. One of our partners on this project calculated that 13 million cumulative days of learning were lost.[2] It wasn't just the missed classes, but it was the uncertainty of when classes would resume. It had a ripple effect on families and the entire community. We knew that reliable electricity was the missing ingredient and that a microgrid at the school would greatly accelerate the entire community's healing after these tragic natural disasters.

Like many early microgrids, the school's project was built primarily for resiliency—to ensure electricity kept flowing no matter what. But microgrids like the one at Pedro Fernandez are not just modern versions of expensive backup generators, they are miniature power plants that operate continuously. These independent, onsite electric generation systems are part of a new community-centered movement I call *local energy*. This new approach to electrification, whether batteries are included or not, offers substantial benefits well beyond backup power. These projects are also cheaper than electricity from the grid. The cost of solar panels and batteries have fallen so far that the Pedro Fernandez microgrid actually reduces the school's electric bills. The microgrid electricity is also much cleaner, particularly in Puerto Rico, where the grid is powered by coal and petroleum—making it one of the dirtiest in the US. Finally, most of the people who installed the microgrid call Puerto Rico home, supporting good-paying, local jobs.

This is where the local energy revolution began

Hurricane Maria's Category 4 winds destroyed nearly everything in their path, leaving 80% of Puerto Rico's grid in tatters.[3] It was no longer feasible for the island's government to defer to the wishes of the island's electric utility. On October 16, twenty-six days after the storm, Governor Ricardo Rosselló issued an executive order declaring that any new solar microgrid could skip the lengthy permitting process and would automatically be approved.[4] This was a watershed moment. First, he was acknowledging that the utility was not close to getting the full grid working again (only 14% was restored by then[5]). Second, and most important, he was acknowledging that solar, batteries, and local energy were ready for prime time.

The governor had removed a major bureaucratic barrier to local energy, and individuals and communities quickly took matters into their own hands. Thousands of small solar microgrids—technologies that had become economically viable only in the last two or three years—were built across the island. House by house and building by building, electricity began to flow again. Resilient Power Puerto Rico (RPPR), a nonprofit group led by architects in New York and Puerto Rico, quickly shipped solar panels and batteries to the island.[6]

Within 30 days after Maria, RPPR had built three solar microgrids for community centers in the poorest parts of San Juan. In that same month, Tesla built a solar microgrid in the parking lot of San Juan Children's Hospital, restoring the hospital's operations to full capacity. Over the course of the first year, Bloomberg reported 10,000 small systems were built.[7]

Companies from Blue Planet Energy to Germany's Sonnen to Pika Energy flocked to the island to install their products where they were needed most. Sunnova, the leading residential rooftop solar provider in Puerto Rico, shifted its business focus from solar-only to solar+battery and has since installed thousands of systems.

A 2021 report by two nonprofit research groups found that 75% of Puerto Rico's total electricity needs can be met by local energy, and for less money than planned investments in the island's Big Grid.[8]

Puerto Rico has become an early leader in what is becoming a global revolution. Hundreds of thousands of small solar and battery systems have been installed across the world as of 2020, with research firm Wood Mackenzie forecasting the number of these systems will grow six times by 2025![9] As we wrapped up our visit to the Pedro Fernandez school, one of the teachers pulled me aside to tell me what a difference the microgrid had already made. Even in the best of conditions, he pointed out, the island's utility struggles to provide consistent electricity, especially to rural areas like Naranjito. In fact, he went on, the community's power had gone out for several hours just the previous week. Thanks to the microgrid, they no longer had to send the children home during outages. It was an important reminder that behind the gee-whiz technology of the new system, at the end of the day, the new microgrid served a more fundamental and humble purpose: to improve the lives of the people of Naranjito by keeping the school running. Someday it will be as unremarkable as the refrigerators and air conditioners we take for granted today.

〔▥〕〔▥〕〔▥〕

This book is about local energy. It is about individuals, communities, and local businesses generating their own energy. It is about choice and fair markets. It is about unleashing innovation in our outdated electric grid. It is about all of us finally taking control of one of the most essential parts of our lives—energy.

Local energy can take many forms. Microgrids, which typically power a building, are a particularly visible and advanced incarnation of local energy. Mini-grids, their larger cousins, can power multiple buildings or homes at once. These systems are popping up on campuses, in neighborhoods, on islands and military outposts, and in public safety buildings. They are even powering entire towns.

Local energy encompasses traditional rooftop solar, both on homes and buildings. When a battery is included, these are often referred to as *solar+battery*. Local energy systems with batteries or backup generators are extremely reliable. When the grid goes down, these systems can operate autonomously. This is called *islanding*, which ensures uninterrupted power to homes and buildings.

For people who cannot access their roofs, or are unable to finance a full system, local energy includes shared solar projects called *community solar*. A particularly impactful form of local energy is the small solar and battery systems that are providing the first electricity to families in rural Africa and India.

Most of us will still rely on traditional grids for years to come, but that does not make local energy less revolutionary. It is already reinventing the electricity industry, offering choice, innovation, and reliability after a century of strict control from electric monopolies. Local energy may not be a panacea, but it is most definitely the future of energy and, as I will explain throughout book, it will improve billions of lives and create hundreds of billions of dollars in opportunities.

The first three chapters of the book dive deeply into the nature of local energy, explaining how it works and why it is urgently needed. Chapter 4, "From Fuels to Technologies," Chapter 5, "Hidden Patterns of Innovation," and Chapter 6, "Billion Dollar Disruptions," cover the technologies, business models, and huge opportunities embodied in the local energy revolution. The seventh and eighth chapters, "Utilities vs the Future" and "The Battle for Public Opinion," explain the ways incumbent electric utilities are erecting obstacles to slow or stop the adoption of local energy. These can slow progress but will ultimately fail. The final two chapters, "Unlocking Our Power" and "Powered by Innovators," reveal how innovators can overcome these hurdles to embrace local energy as individuals, communities, investors, and startups.

As I put this book together, I made the tough decision to limit the discussion of greenhouse gas emissions and climate change, factors that many attribute to the increasing frequency and magnitude of destructive weather events like Hurricane Maria. These are urgent issues, but they are also well covered. By my count there are more than 500 books already focused primarily on this topic. Climate change is a divisive topic in many communities, and that controversy risks distracting from the core point of this book: *clean, local energy systems like solar+battery are becoming a cheaper source of electricity than our centralized fossil-fuel powered grid, regardless of subsidies.*

This book is about the practical business side of clean energy and the enormous opportunities for innovators. It offers a guide for accelerating the transition to clean energy by embracing technology and putting customers first. Rather than another top-down, prescriptive policy-first approach, I offer you a faster, more personal path to a clean energy future. The plan is simple: unleash and tap into the incredible talents and energies of innovators. This requires tens of thousands of entrepreneurs, policymakers, and investors from across the world to join the movement. It also requires hundreds of thousands of everyday innovators—people putting solar on their roofs, opting into community solar programs, buying electric vehicles, or just sharing what they have learned with elected officials and community leaders. *It is for these people that I decided to write this book.*

So, let us begin. Like all big stories, the best place to start is at the beginning.

THE BIG GRID'S GLORIOUS BEGINNINGS

Electricity is a nearly perfect form of energy. It can be instantly distributed across hundreds of miles, requiring little more than thin strands of insulated copper. It can be converted to light, heat, and motion more efficiently than any other type of energy. Consuming electricity generates no pollution or waste. Tiny bits of electricity power the billions of microscopic transistors in modern electronics. Giant flows of electricity power factories and cities. Most of us give little thought to electricity, yet it silently and invisibly powers the very foundations of our modern society. One can see why, after electricity was first harnessed for human use, it captured the cultural imagination. An inscription from the early 1900s on Washington, DC's Union Station, expresses the awe felt about a technology we now mostly take for granted.

ELECTRICITY—CARRIER OF LIGHT AND POWER
DEVOURER OF TIME AND SPACE—BEARER
OF HUMAN SPEECH OVER LAND AND SEA
GREATEST SERVANT OF MAN—ITSELF UNKNOWN

Thomas Edison is credited with inventing many of society's formative technologies like the practical electric lightbulb, movies, and audio recording. But he gets little credit for what I believe was his greatest invention of all. On September 4, 1882, Edison pulled a giant switch at his Pearl Street Station plant and powered up the world's first electric grid. Instantly, a square mile of lower Manhattan lit up with the clearest, safest, and most affordable lighting humankind had ever experienced.

What makes this first grid so remarkable was not its technology. Most of it had existed in one form or another for years. Instead, it was the vision to weave together existing innovations like lightbulbs, insulated wires, power meters, and large coal generators into one large system that simultaneously served hundreds of customers. In my opinion, Edison's greatest invention was *centralized electric power generation* and a business model that made electricity afford-able for the masses. "We will make electric light so cheap that only the wealthy can afford to burn candles," Edison exclaimed at the time. Years before Henry Ford's pioneering assembly lines, Edison was arguably one of the first businesspeople to capitalize on the awesome power of *economies of scale*. For the glorious century that followed, bigger was better—and cheaper—every time.

The expansion of electric grids from lower Manhattan to entire continents is one of the most riveting stories in business history. The "current wars" between Edison's *direct current* (DC) technol-ogy and the competing *alternating current* (AC) commercialized by Nikola Tesla and his benefactor, George Westinghouse, have been memorialized in numerous books and even a recent movie by the same name (see the Appendix for a primer on electricity). But it was another electric battle in the industry's early days, fought by a man named Samuel Insull, that is most relevant to this book.

Insull began his career as Edison's secretary and rose to become one of the most powerful and wealthy people in America. His major contribution to our electricity system was convincing a public still reeling from the abuses of the railroad and steel trusts that electric-ity was an exception. He argued that electric power was a *natural monopoly* because competition undermined access to the vast quan-tities of capital required to construct ever-larger power plants. He convinced the country that electricity was best supplied by a small handful of businesses, each granted exclusive franchises in their geographic regions. Samuel Insull created the *regulated monopoly*, a business model so successful that it has thrived for a hundred years, largely unchanged and unchallenged.

Edison, Tesla, Westinghouse, and Insull are the founding fathers of one of the largest industries on earth—what I call the *Big Grid*.[10] Their vision, audacity, financial engineering, and occasional dirty tricks created the largest, most sophisticated machine every built—the electric grid. It connects hundreds of gargantuan, centralized power plants across millions of miles of power lines to hundreds of millions of homes, businesses, and industries of every kind.

It is difficult to overstate the impact of electrification on our culture. In many ways, electricity powered the ascent of the middle class. Lighting extended peoples' days. Conveniences like washing machines and electric ovens freed up countless hours of labor. Millions of people suddenly had leisure time for another electric transformation, broadcast radio. As acclaimed author and one of my favorite thinkers, Vaclav Smil, points out in his book *Energy and Civilization*, "One of electricity's most consequential social impacts has been to transform many chores of household work and hence to disproportionately benefit women."

Electricity made factories cleaner, safer, and more efficient. Electric trolleys made cities more accessible. Elevators made high-rise buildings possible. Electricity revolutionized the manufacture of steel and aluminum. President Roosevelt's crusade to electrify rural America in the 1930s completed the puzzle and made the US an electric powerhouse. It is no wonder, then, that the National Academy of Engineering ranked electrification as the most important engineering achievement of the 20th century, ahead of airplanes, automobiles, telephones, and the internet. The Academy's recognition of the grid captures its magnificence:

> Scores of times each day, with the merest flick of a finger, each one of us taps into vast sources of energy—deep veins of coal and great reservoirs of oil, sweeping winds and rushing waters, the hidden power of the atom and the radiance of the sun itself—all transformed into electricity, the workhorse of the modern world.[11]

But as would be true for any 135-year-old, continuously operating piece of machinery, the Big Grid is also a bit of a mess—an inefficient, jerry-built patchwork of outdated technologies managed by a Gordian knot of arcane regulations and entities. Like a magnificent gothic cathedral, one marvels not only at its vast scale but at the fact that it still functions. In our modern digital age, the Big Grid remains stubbornly rooted in Edison's original design—a single, sprawling analog circuit. It is desperately overdue for an upgrade. But change is proving to be incredibly difficult. It faces the inertia of trillions of dollars of aging assets and the active resistance of one of the most powerful and politically influential industries in the US—the monopoly electric utilities.

It would be easy for us to sit back and let the politicians and giant corporations apply their cumbersome, risk-averse processes to fix the grid. But we do not have to. Edison had a vision to aggressively embrace existing technologies that revolutionized energy in the 19th century. We can do the same, using the tried-and-true technologies available today. Not only that, reimagining and upgrading our vast electricity system at a more local scale will unleash a surge of innovation. This offers unprecedented opportunities for innovators and investors, as well as widespread economic benefits to individuals and communities around the globe.

THE BIGGEST BUSINESS OPPORTUNITY IN HISTORY

Many of the world's most inspiring leaps in progress occur when governments, private business, and thousands of people align with large sums of capital toward truly grand challenges. The investments required can be colossal, but the gains, both societally and financially, can be even bigger. For example, the US highway system is often held up as the largest US infrastructure project in history. From 1956 through 1995, the government invested $549 billion in the US highway system.[12] These roads opened the country, revolutionized commerce, and have paid for themselves many times over. NASA's Apollo missions put 12 men on the surface of the moon for a price tag of $120 billion (in 2020 dollars).[13] The accomplishment inspired the world and spun out countless innovations, from aircraft anti-icing systems to scratch-resistant lenses. From 2000 to 2020, a staggering $1 trillion of venture capital was invested, creating much of the modern internet, media, and communications technology we know today.[14]

Projects like these not only lift and inspire our society, they also create hundreds of thousands of jobs, commercialize world-changing technologies, and provide substantial opportunities for investors willing to make bets on the future. But all of these are small in comparison to what is coming.

The world is embarking on one of the most important projects in its history—the transition away from fossil fuels toward renewable energy. Across the board, experts forecast unprecedented investments in the coming decades. Goldman Sachs predicts $16 trillion could be invested through 2030 in renewable energy infrastructure.[15] US Energy Secretary Jennifer Granholm's estimate is even higher: a $23 trillion market by 2030.[16]

The scale of these investments is breathtaking. Remember, these figures represent *investments*, not handouts, write-offs, or expenses. For every trillion dollars put into renewable energy, investors will receive even more money back over time. Of course, the returns for the planet's future are priceless. If the governments of the United States, the European Union, China, and India decide to step up and more aggressively invest to reduce greenhouse gas emissions, these numbers will be 50%–100% higher.

To help get your head around the scale of these numbers, consider what it will take to put $10 trillion dollars to work. Over 20 years, investing this money will require thousands of companies, each investing hundreds of millions of dollars every single year for decades. This means millions of new jobs, tens of thousands of startups, and economic prosperity spread across the US and the world.

The most exciting part for me is that some portion of this investment will go toward local energy. The final amount is up to us, but in every case, investments in local energy will accelerate the adoption of renewable energy, and supercharge innovation, entrepreneurship, and investor returns. Best of all, with local energy, every homeowner and building owner can become a small entrepreneur—buying and selling electricity in a community marketplace of kilowatt hours, electric storage, and resiliency. Local energy shifts the profits from these enormous investments toward individuals and communities, away from Wall Street and the shareholders of giant corporations. Whether in a small farm in rural Kenya or apartments in New York City, local energy *democratizes* electricity, entrepreneurship, financial independence, and self-reliance.

If the renewable energy industry has a North Star, it is a man named Amory Lovins, the cofounder of the deeply respected "think, do, scale tank" RMI (recently renamed from Rocky Mountain Institute) This is the same firm that was behind the Pedro Fernandez school project. Amory is a personal hero of mine, and I will share his insights throughout the book. Ever eloquent and provocative, he described the business opportunity in a 2012 TED talk:

> I've described not just a once-in-a-civilization business opportunity, but one of the most profound transitions in the history of our species. We humans are inventing a new fire, not dug from below, but flowing from above; not scarce, but bountiful; not local*, but everywhere; not transient, but

permanent; not costly, but free.[17] (*Author's note: in this context, "local" refers not to scale but to a limited set of locations.)

RMI has grown tremendously since it started in Amory's house in 1982. With offices around the world and hundreds of brilliant people, the organization has been at the forefront of the global transition to clean energy. In many ways, the mission behind Freeing Energy was born from the passion and optimism of Amory and his many colleagues who have collaborated with me over the years. RMI has inspired me and many, many others.

WHAT IS FREEING ENERGY?

Freeing Energy is a mission that has grown into a website, a podcast, and most immediately, a book. Before I answer what Freeing Energy really means, I would like to share how I first came to ask the question.

My journey

My first step toward writing this book took place on the afternoon of May 15, 2014. A software company I led for years had just announced it was being acquired by IBM. It was a momentous day. As the world's business media was buzzing about the deal, I stepped away for a few moments to take it all in. Unexpectedly, my excitement was overshadowed by a sobering question. Here I was with years of experience and a comfortable nest egg for myself and my family, so what was I going to do with it? The most obvious path was to become a venture capital investor again. Or I could build another company, or even retire. But then I thought of all the people who sacrifice so much every day to make the world a better place. How could I do any less? Over the next few months, my wife and I decided to give away a large portion of the money we had made, leaving just enough for me to get started on a new mission. My journey had begun.

The next step took place inside a mud hut. It was the summer of 2015, and my family was visiting Africa. While we were there, we wanted to meet and learn from other families whose lives were completely different than ours. We decided to visit a Samburu village deep in the bush of Kenya. To get there, we had flown for two hours in a tiny propeller plane, then driven off-road another two hours.

The journey was worth it. As we sat around an earthen home lit only by a small fire in its center, a young Samburu mother showed us how she prepared her family's meals. The smoke burned my eyes, and I found it increasingly difficult to breath. Through the translator, I asked her how her children dealt with the harsh air. She told us they accepted it because there was no other way to light her home. I remember thinking there had to be a better way.

It was the final step in my journey that defined Freeing Energy. It was early 2016 and I was now IBM's VP of Corporate Strategy. I was working out of the company's headquarters in New York. My team was looking at a wide range of industries to see where IBM's digital technologies could make the biggest impact. I was surprised to learn that the electricity industry, also called the *power industry*, was still operating with the same basic technology architecture that Thomas Edison invented 140 years ago. Furthermore, during my research, I stumbled onto something much bigger and more exciting; something far outside of IBM's focus. Solar power was on the verge of becoming cheaper than coal, natural gas, nuclear, and even wind. Batteries were following similar, stunning cost declines. This industry was about to be turned upside down. A trillion-dollar disruption was underway.

Then it all clicked. Clean energy was my future. By shifting my career, I could participate in three world-changing mega-opportunities at the same time: reversing the damage our modern energy systems are having on the environment, helping nearly a billion people kick-start the economic development that comes from electrification, and participating in one of the biggest business opportunities in history. I realized that solar and batteries were the single technology building block that was fueling all three of these. I had to be part of the clean energy industry. The technology and policy of solar and batteries was my new mission.

But I was an outsider. How could I possibly hope to get up to speed and meet the movers and shakers? More importantly, knowing my own efforts were dwarfed by the magnitude of the challenges, how could I help thousands of other people become as excited as I was?

So, *Freeing Energy*, the book, was born. I left IBM in early 2017 and began my research.

As I interviewed experts, two discoveries reinforced my belief that these new energy technologies would be a vital part of our future electricity system. I learned that the grid is fragile, frighteningly so. Catastrophic grid failures in Puerto Rico, California, and Texas exposed the glaring weaknesses stemming from years of under-investment and stagnation.

This led me to the second discovery: The electricity industry was nothing like any business I had ever seen. Utilities were granted exclusive monopolies by state governments and faced virtually no competition. Their profits are guaranteed by the government. Consumers and businesses have virtually no choices or alternatives. It was no wonder that failures are tolerated, environmental impacts are overlooked, and prices are going up.

I had spent a career building companies that brought new technologies into tired legacy industries. I saw firsthand how disruptions broke the incumbent logjams and democratized outdated industries, delivering better, cheaper solutions to millions of customers.

In high school, I built a company that brought real-time graphics to early text-only microcomputers. In the early '90s, I led a company that sold one of the first network applications for personal computers. During a brief stint at the world's top strategy consulting firm, I developed complex economic models for a giant telecommunications company as that industry deregulated. I was a venture capitalist in the early years of the internet, working directly with the people and companies that built the foundation of today's internet. In the late '90s, I ran one of the largest internet consulting firms in the world, helping set the strategies and building the systems that helped our Fortune 500 clients to open for business on the internet.

These unique experiences have provided me a ringside seat to the patterns of innovation and disruption wrought by new technologies. I have seen this from both the perspective of an entrepreneurial CEO as well as a Fortune 500 strategist. The clean energy revolution will be larger than anything I have been a part of in my career. But for this opportunity to be realized with the urgency it demands, it requires far more than anything I could do on my own. It requires tens of thousands of innovators to join the movement and quickly begin making impacts. *It is for these people that I decided to write this book.*

After I dusted off my notes from my undergraduate electrical engineering classes, I got out my passport and began to travel the globe, where the future of clean, local energy was already underway. I visited enormous solar panel factories in China, microgrid projects across the US, startups in Africa, and even ventured to the top of a wind turbine outside Denver. I met with and interviewed over 300 experts, from scientists to politicians, to understand the role that local energy is playing in the transition to a clean energy future. What I discovered was a movement that transcended geography and ideology. Patterns emerged that foretell which organizations and which technologies will prevail and which will be rendered irrelevant. Most importantly, I met and was inspired by visionary and committed innovators, advocates, and entrepreneurs who are devoting their lives to one of the most important opportunities in the 21st century.

Let me briefly introduce you to a few of them. Steph Speirs, the CEO of Solstice, is delivering the promise of solar to low-income families in the US that could not otherwise afford to finance their systems. Mac McQuown, a retired banker, has built one of the most sophisticated microgrids in the world on his farm in California. Andrew "Birchy" Birch, the CEO of OpenSolar, helped pull together a broad coalition of businesses and governments to drive down one of the most expensive parts of solar installations. Bill Gross, the famous internet entrepreneur, has shifted his focus to climate tech and has started two of the most exciting clean energy companies in the industry. Samir Ibrahim, the CEO of SunCulture, led his team to create the first affordable solar-powered irrigation system for smallholder farmers in Africa. The late Jim Rogers, who was the CEO of utility giant Duke Energy, was one of earliest leaders in his industry to embrace clean energy. This is just a small sample of the dozens of visionaries whose stories I have chronicled throughout the book.

Why now?

We are living in a crucial moment in human history. One of our most essential commodities—electricity—is poisoning our planet, impoverishing a billion people, and is at an ever-increasing risk of failing the rest of us. At the same time, decades of brilliant innovations have gifted us with solutions that offer cheaper, cleaner electricity to everyone on the planet. The gap between these two potentials is why this is one of the biggest investment opportunities in business history.

Two areas of innovation are needed to realize this opportunity; the rest of the book will explore both in depth. The first is technology. Solar cells, batteries, and electronics are the "LEGO bricks" of our energy future. We can use one of these "bricks" to make a solar lantern. A few thousand can lower the costs and increase the resilience of a suburban home. Several million will power a city. No other source of energy has ever been as flexible and scalable.

Second, we need to "free" energy. We need to free it from the tyranny of dirty fossil fuels. We need to free it from the outdated business model of regulated electric monopolies. We need to free energy from a century of technology stagnation. But, most important, we need to free energy so that millions of families and small businesses can have real choice, taking control of one of the most important foundations of their daily lives.

What I am suggesting is not a small-scale intervention. Local energy needs to shed its past as a niche industry that sits quietly in the shadow of the Big Grid. If we are going to unleash the creative potential of innovators and empower billions of people, it is going to require a bottom-up and outside-in reinvention of the very architecture and business models of our electricity systems. So why am I convinced this outside-in approach is the right thing to do? Because the business model and oversight of today's Big Grid is failing us. The scale of the problems and the magnitude of the risks is growing faster than utilities and regulators can keep up with.

The next chapter will explain the problems with the Big Grid and why it is increasingly urgent that we find more ways to address them.

CHAPTER 2

YOUR POWER IS FAILING

On December 22, 2008, shortly after midnight, not far from a coal plant in Kingston, Tennessee, a six-story earthen dam collapsed and a nightmare was unleashed.[1] Over the next hour, 1.1 billion gallons of a dark sludge called coal ash slurry poured through the breach, spreading through the landscape like the alien from the 1950s sci-fi horror film, *The Blob.* By the time this wave of toxic goo stopped, it had destroyed twelve homes and blanketed more than 300 acres of nearby land.[2] Lead levels downstream from the spill spiked to 400 times the Environmental Protection Agency's (EPA) recommended safe limit. The spill released 180,000 pounds of arsenic, more than twice the amount from *all* US power plants the prior year.[3] Perhaps due to the absence of heart-tugging images of petroleum-soaked seabirds and spoiled beaches, the Kingston spill received far less attention than other energy-related environmental disasters, like the 1989 Exxon *Valdez* spill. But in volume, the Kingston spill was 40 times worse than the *Valdez,* making it the largest industrial spill in US history.[4]

FIGURE 2.1 *Homes devastated from the coal ash spill in Kingston, Tennessee. Source: Rick Ray/Shutterstock.com.*

For fifty years, the Kingston plant had been burning about 14,000 tons of coal every single day. And every lump that had been incinerated left behind a fine particulate ash laced with heavy metals like mercury, lead, arsenic, chromium, and uranium. To keep it from blowing away, it was mixed with water and dumped into a pond just outside the coal plant. This was the toxic brew that washed over the local community.

When I first read about the spill, years after it occurred, I was incredulous. How could I have not heard about this? Surely, the health risks of coal ash and this spill were overstated. But judging from the fate of many of the workers hired to clean up the Kingston spill, it may be deadly. The local newspaper has reported that ten years after the original 900-person team cleaned the spill, more than 30 had died and 250 became chronically ill.[5] Many of these workers have sued. As of 2020, the case was still working its way through the courts. The utility that owned the coal plant and the firm contracted to clean up the spill both deny any wrongdoing or culpability.

Here is the crazy part. This is just one single site.[6] According to the EPA, there are 735 of these enormous coal ash "ponds" across the US.[7] *Less than 10% of 735 coal ash ponds in the United States* have linings that prevent the toxins from leaching into the soil and water.[8]

So it is not surprising that data submitted to the EPA and analyzed by an environmental group found that 91% of sites reporting groundwater tests had toxic substance levels exceeding federal safe standards. Coal ash is the largest industrial waste product in the US, and more than 100 million tons of it is left behind each year. According to the American Coal Ash Association, "The quantity of coal combustion products in disposal across the US exceeds 2 billion tons."[9] This is enough to cover all of New York City in 9 feet of coal ash.[10]

Most of us are aware that our electricity system is harming the environment. But few really understand the range of damage because it is spread across thousands of plants across the country that together add up to something so great, it exceeds our ability to fully comprehend it.

The stakes involved in reimagining our electricity system are similarly large, and they involve more than just the environment. In this chapter I will examine five of the main reasons we must act quickly and decisively to change the way we generate and distribute electricity. They are not only about averting ongoing environmental and health catastrophes—although that is a major part. And it is not just because the current laws and regulations are outdated and forcing families to spend billions of dollars propping up an uncompetitive industry. It is also because it will allow more people all over the planet to enjoy the benefits of electricity, and change billions of people's lives for the better.

POISONING THE PLANET (#1)

The coal ash released in the Kingston disaster is only one type of pollution from coal power. No other type of power plant emits more greenhouse gases. Half of all the mercury released into the environment comes from coal plants.[11] A Harvard study found that the particulate pollution from burning coal, diesel, and gasoline led to 10 million premature deaths globally *in a single year.*[12] None of this includes the environmental damage from mining. Mountaintop removal and strip-mining coal have impacted 18,000 square miles of US land, the size of ten Rhode Islands.[13]

FIGURE 2.2 *Mountaintop removal coal mine in Blair County, West Virginia.* Credit: Orjan Ellingvag/Alamy Stock Photo.

If there is a silver lining, it is that coal is very much on the decline. From its peak in 1988 when it powered 57% of the US grid, coal power has fallen to 20% in 2020. Experts predict the decline will continue until coal is phased out entirely. The irony is that cheap natural gas, more than coal's environmental impact, has driven this decline. But even after coal- powered electricity is history, our children and grandchildren will be stuck paying hundreds of billions of dollars in cleanup costs for ash ponds, strip mines, and other scars from the era of "cheap" coal. The CEO of the utility that operated the Kingston plant put it well: ". . . the industry is now dealing with a hundred years of deferred costs."[14]

It is not just coal

I started with coal because it is, admittedly, the bad boy of power industry pollution. But every type of power generation impacts the environment in some way. Even solar panels and wind turbines need recycling or disposal when they are retired, although it is a fraction of the scale and far less toxic than coal waste (I present the data on this in Chapter 8, "The Battle for Public Opinion.") What I learned about coal, nuclear, natural gas, and biomass was so surprising I had to ask myself if perhaps the heavy focus on greenhouse gas emissions is deflecting attention from the equally threatening and immediate impact of heavy metals, particulate air pollution, and toxic wastes.

Nuclear power. Nuclear power accounts for 20% of the US grid, a number that has remained largely unchanged for decades. As I explain in Chapter 8, nuclear power is one of the most contentious forms of electricity generation. Its proponents boast that nuclear plants emit no greenhouse gas emissions, and the volume of physical waste is tiny. But, as its critics will point out, that waste is very toxic, and its comparatively tiny size means it is particularly concentrated.

The US requires nuclear waste to be securely stored for 10,000 years. Some European governments require 100,000 years. For most of us, those numbers are too abstract to mean anything. In practice, it is nearly impossible to plan for 100 years, let alone 100,000. Consider that the oldest Egyptian pyramids were built 5,000 years ago,[15] and it was just 20,000 years ago, during the last glacial period, that Canada and parts of the US were covered in ice.[16]

Most governments, including the US, have been paralyzed by the scope and implications of long-term storage and have no plans in place. In the meantime, most of the world's nuclear waste is stored in giant casks that sit outside of the plants that created it. There are 90,000 metric tons of nuclear waste stored from US nuclear power plants and more than 250,000 metric tons globally.[17] While the US government has set aside some funds for long-term storage, experts agree it is not nearly enough. The long-term costs of keeping nuclear waste safely out of the environment and out of the hands of terrorists will fall to thousands of generations after us.

Natural gas. No other type of power plant has expanded its footprint on the US grid as quickly as natural gas. From 14% in 2000, it has become the country's largest source of electricity in 2020 at 39%. It produces no physical waste, and its greenhouse gas impact is smaller than coal. Oh, and it is also incredibly cheap thanks in part to *fracking*, which has extricated huge volumes of low-cost gas. Fracking, however, uses millions of gallons of water as well as largely unregulated fracking fluids to force the reluctant gas out from deep inside shale rocks. These fluids can leak into water tables, and they are not always disposed of properly when drilling is complete. While burning natural gas produces less CO_2 than does burning coal, the main component of gas, methane, is an even more potent greenhouse gas. Large volumes are leaked during drilling and distribution, although lax regulations leave precise amounts unclear, and estimates vary widely.

The industry has another growing problem—zombie wells. Many of these inactive, uncapped wells are leaking methane. The pandemic accelerated the bankruptcies of oil and gas companies that are technically responsible for decommissioning these wells. The think tank Carbon Tracker estimates up to $280 billion will be required to properly retire them all. Public data indicates that states have less than 1% of that money set aside.[18] Once again, these costs will fall to our children and grandchildren. But whatever critiques natural gas deserves, it is environmentally preferable to coal.

Biomass. You have to squint your eyes and tilt your head to argue that cutting down trees and burning them is a form of clean energy. Yes, regrowing trees recaptures much of the CO2 emitted when they are burned, but for the 20 to 100 years it takes for the trees to grow back and reabsorb the CO2, the environmental footprint of biomass looks an awful lot like that of coal. In fact, the world's largest bio-mass plant, the Ironbridge plant in the UK, was a converted coal plant. Both biomass and coal create prodigious amounts of CO2, and both leave behind a fine ash (although biomass ash has fewer heavy metals because it comes from trees, not from the ground). Biomass powered electricity is very expensive, so it is unlikely it will expand its footprint beyond 0.3% of the US grid in 2020.[19]

All these costs do not even include the harder-to-estimate health-care costs as fossil fuel pollution makes its way into our air, water, and ultimately our bodies. Nor do they include the much larger indirect costs contemplated by most of the models analyzing greenhouse gas impacts on weather and climate.

What can be done?

Can your local energy project really make a difference? Yes, it can. Assume a solar rooftop for an average-size home generates one megawatt hour (MWh) of electricity per month. If your solar MWh is replacing coal, you are keeping 1,100 pounds of coal in the ground each month and eliminating 2,100 pounds of CO2 and 185 pounds of ash from ever being created. If your solar offsets a natural gas plant, you would keep the equivalent of 1,400 pounds of CO2 out of the atmosphere.

FRIGHTENINGLY FRAGILE (#2)

The amazing fact about power outages is not how many we have but that we have so few. Somewhat ironically, it is the sheer scale of the Big Grid that makes it so vulnerable. Nearly all the grid failures—about 90%—are due to failures in the wires that connect your home to distant power plants.[20] Those miles and miles of wires are like a giant chain. If any link fails, the entire chain fails with it. Engineers would say the system is riddled with *single points of failure*, meaning if any part of the system breaks, a disproportionately larger part of the system breaks with it. In some ways, it is like a giant extension cord from the power plant to your house—if the cord breaks anywhere along its length, your house goes dark.

To the credit of electric utilities, they do an impressive job of limiting the duration and scope of outages. In particularly vulnerable locations, they can reinforce wires, poles, and components. If more reliability is needed, they can bury power lines underground, though installation and maintenance can be expensive. For critical facilities, they can install several independent sets of wires. But often their best strategy is to simply detect and repair broken lines as quickly as possible. I have met some of the linemen who do this. They get out of bed at 3:00 a.m., drive for miles, and repair wires in the face of ferocious winds, blinding rain, and freezing temperatures. This is an incredibly tough job. We owe them our deep respect and gratitude.

One of the biggest surprises in my research was just how vulnerable the Big Grid has become. Outages are increasing in frequency, but it was the potential for much larger and longer outages that left me most anxious. We cannot live without the Big Grid for the foreseeable future, but due to any number of risks that could cause widespread failure, several of which are listed below, our electric future cannot continue to rely on it alone.

The Big Grid is a convenient target for Mother Nature

Mother Nature has an arsenal of ways to cut power to your house. Every time a tree comes in contact with a power line, it can short the electricity to ground and trip circuit breakers. Trees can fall on lines, wind can blow lines into trees, and hot weather can cause lines to sag and touch trees. High winds and ice storms can sever power lines. Earthquakes and floods can take out every part of the Big Grid. Heat waves and droughts can dry up the water needed to cool coal, natural gas, and nuclear plants, forcing them to shut down.

You might be surprised to learn that squirrels are a common cause of smaller outages. These curious rodents inadvertently, if very briefly, find themselves touching two different circuits at once, tripping breakers, and creating more than 3,000 small outages in 2016 alone, according to the American Public Power Association's Squirrel Index. (I am not making this up.)[21] Reading through these reports, you see classic utility-style commentary like "The outage at 8:37 a.m. was caused by a squirrel. The animal did not survive the incident."

Of course, large-scale outages that involve thousands of homes are a far more serious concern, and the number of weather-related blackouts is rising.

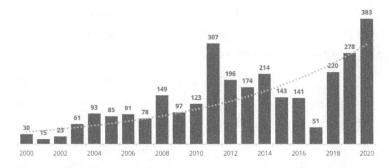

FIGURE 2.3 *Weather-related outages are increasing. Source: US Department of Energy, Form OE-417 (freeingenergy.com/g206).*

In her book *The Grid: The Fraying Wires Between Americans and Our Energy Future,* Gretchen Bakke describes a major outage that occurred on August 14, 2003:

> . . . the largest blackout in our nation's history [at the time], and the third largest ever in the world, swept across the eastern half of the United States and parts of Canada, blacking out eight states and 50 million people for two days. The blackout, which covered 93,000 square miles, accounted for $6 billion of lost business revenue. If ever it was in doubt, the 2003 blackout proved that at its core America's economy is inexorably, indubitably electric.

The cause? Several overheated power lines sagged and shorted against trees that had not been pruned back. This was compounded by a computer bug at a local utility in Ohio delaying alerts to operators for over an hour, preventing them from taking immediate action when it was required.[22]

This outage was a painful lesson that, under the right (or wrong) circumstances, the highly interdependent components of the grid can fall like dominoes. Countless upgrades and improvements have been made since, and this type of failure is unlikely to happen again. But as the entire state of Texas painfully learned in early 2021, some outages are not due only to failed power lines. Power plants themselves can go offline—sometimes many at once.

The Texas grid failed after an unprecedented cold snap hit the state in mid-February. It was far colder and lasted much longer than the state's regulatory planners had ever modeled. The Texas grid was unprepared. While some politicians singled out frozen wind turbines, the largest failures were in fossil fuel plants. Coal stocks were frozen. Natural gas valves became stuck. A nuclear plant's water supply froze.[23] All affected plants were taken offline, and because Texas's grid was independent from the grids that serve the rest of the United States, it was unable to draw electricity from nearby regions. According to the head of the Electric Reliability Council of Texas (ERCOT), the state's regulatory body, the grid "was seconds and minutes" from complete collapse. Texas came precariously close to a state-wide outage lasting weeks. As it was, 4 million went without power, some for several days, and more than 200 died from the multi-day freeze.[24]

As we've seen in Puerto Rico and Texas, many grid failures involve natural disasters that shut down the electricity system. But we can also see grids inflicting great damage on nature. California's recent fires are an example of this tragic twist on the usual narrative. Many of the state's enormous wildfires have been sparked by poorly maintained power lines. This is not a new problem, but on November 8, 2018, it took on new and horrifying proportions.

The Camp Fire burned 240 square miles, destroyed more than 18,000 structures, caused $16.5 billion in damages, and directly led to the deaths of 84 people. This was a final straw for regional utility, Pacific Gas and Electric (PG&E). Facing a torrent of lawsuits, the company declared bankruptcy and pled guilty to 84 counts of involuntary manslaughter, ultimately paying out more than $13 billion to the victims.[25]

To limit their power lines from sparking more fires in the future, PG&E and other California electric utilities have since instituted widespread *public safety power shutoffs*. These intentional blackouts are designed to cut off electricity from exposed power lines in areas and during times that are particularly prone to fires. These outages affect hundreds of thousands of people and put extraordinary strain on those dependent on medical equipment and other electric products necessary for their lives and livelihoods.

So, how does the state solve this problem? While it slowly invests billions of dollars and many years in fixing and upgrading its power lines, the state's regulators have ordered deployments for the only solution that is quick, affordable, and safe enough to make an impact now: local energy. In the summer of 2020, the California Public Utilities Commission required the rapid deployment of microgrids and other resiliency projects that will help keep electricity flowing, particularly in critical facilities.[26] This is creating a new California gold rush. Microgrids and residential solar+battery systems are not mere Band-Aids for the state's broken grid, they are its future. Along the way, they are creating jobs, helping clean the environment, shrinking electric bills, and making communities more resilient across the state.

Texans are learning similar lessons. During the 2021 freeze, as well as during several hurricane-triggered outages in previous years, a regional grocery chain, H-E-B, continued to make news as its microgrid-powered stores remained open for business. First responders were able to use the stores for operating bases. Food remained refrigerated and fresh. Customers had a place of safety and shelter if they needed it. While H-E-B's microgrids are primarily powered by natural gas, they show the promise of how local energy can be far more resilient than the Big Grid in the face of Mother Nature's growing impact.

Solar flares and electromagnetic pulse weapons (EMPs)

Of all the things I learned while researching this part of the book, none has unnerved me more than the civilization-ending potential of large-scale electromagnetic (EM) energy. Humans have successfully harnessed this force for centuries, albeit at much smaller scales. It is what allows generators to turn motion into electricity, and it allows motors to turn electricity into motion. At even smaller scales,

EM energy is tamed to carry mobile phone and Wi-Fi signals. For the most part, EM energy has no effect on living things, which is fortunate, because every century or two, one of the giant solar flares routinely emitted by the sun bathes the earth in a massive amount of EM energy. This energy may pass harmlessly through our bodies, but the same cannot be said of our electric systems.

Years before there were electric grids, the US was crisscrossed with wires of another kind—telegraph wires. When the largest solar flare ever recorded hit the earth in 1859, those long wires acted like giant antennas. They transformed the flare's energy into a massive surge of electric current. Telegraph machines caught fire and several buildings burned down. The "Victorian internet" was devastated.

In 1989, when millions of miles of power lines covered the planet, a far smaller solar flare took down the entire grid in the Canadian province of Quebec for 10 hours.[27] Massive solar flares happen more often than you may want to believe. An 1859-scale flare crossed Earth's orbit in 2012.[28] Fortunately, it just missed our planet, which had moved past that point only days before. Had it hit, the damage could have been terrifying. Even scarier, a study in *Nature* estimated that flares 1,000 times larger than the one in 1859 occur every 800 to 5,000 years.[29]

Electromagnetism has also been weaponized. Specialized nuclear bombs have been designed to specifically generate enormous electromagnetic pulses (EMPs), creating an even more damaging version of a solar flare. These EMP weapons are designed to send their targets back into the Stone Age. It is believed that most nuclear countries, including North Korea, have created EMPs that can be launched with conventional missiles and are able to land a few miles off the coast of a targeted country. In testimony to the US Congress in 2013, retired CIA Director James Woolsey described the worst-case consequences of a country wide EMP attack:

There are essentially two estimates on how many people would die from hunger, from starvation, from lack of water, and from social disruption. One estimate is that within a year or so, two-thirds of the United States population would die. The other estimate is that within a year or so, 90 percent of the U.S. population would die. We are talking about total devastation.[30]

This is certainly terrifying, but I should be clear that expert opinions on the impact of large EM bursts vary widely. The truth is, no one knows for sure because there are simply too many unknowns. But the mere possibility demands that we do more than keep our fingers crossed. Fortunately, local energy systems like microgrids, with their much shorter wires and their ability to operate independently from the Big Grid, are more likely to survive. Woolsey called out microgrids specifically as "inherently less susceptible" to EM bursts in an article he co-authored in the *Wall Street Journal*.[31]

Cyber and physical attacks

Two days before Christmas 2015, a quarter million people in the Ukraine lost power for several hours. The cause was a frightening milestone in the history of electricity. Most experts have concluded that it was the first successful, large-scale cyber-attack on electric grid infrastructure. The electronics controlling 30 substations were *bricked*, meaning they were hacked remotely, and their embedded code was overwritten. Most of the devices had to be physically replaced. There is near unanimous agreement that Russia was behind it, though, of course, Russian leaders deny it.

In late 2020, the Indian city of Mumbai went dark, with some homes losing power for 10 hours.[32] Initial reports suggested it was a cyberattack from regional rival China, but India's government has since denied this claim. Herein lies the insidious downside of our always-connected global networks—not only are the origins of these attacks nearly impossible to trace, but we also cannot even be sure they are attacks. It is a new blend of cold wars and terrorism.

Is this really a threat? Can adversaries really shut down grids and other critical infrastructure? Think about it. Dozens of banks with state-of-the-art security systems have been breached in recent years. Is it really possible that the Big Grid, some parts of which still rely on Windows 7 computers, is somehow immune?

Make no mistake, experts concur that countries like the US, Russia, China, even Iran and North Korea, have breached each other's grid control systems, at least to some degree. In June 2021, when Energy Secretary Jennifer Granholm was asked whether US adversaries had the ability to shut down the US grid, she responded, "Yeah, they do."[33] It is small solace that a cyber detente is keeping these countries from initiating what former Defense Secretary Leon Panetta famously calls a "cyber-Pearl Harbor"—at least for now.

Cyberattacks do not even have to directly target infrastructure to be effective. In the spring of 2021, Colonial Pipeline, the company operating one of the largest petroleum pipelines in the US, was the victim of a ransomware attack. Rather than targeting the pipeline's physical control systems, the attackers disabled the company's business computer systems. Since this pipeline delivers almost half of the gasoline and diesel fuel for the US East Coast, the company decided to shut down the pipeline for six days while it confirmed the attack had not inserted malware into the computers controlling the pipeline itself.

The most successful attacks will most likely be a combination of cyber and physical attacks. Most of the US grid infrastructure has little more than fences to thwart attackers. The US federal group that oversees the country's grids is called the Federal Energy Regulatory

Commission, or FERC. In 2015, a leaked internal FERC report painted a dire portrait of the grid's vulnerability.[34]

> Destroy nine interconnection substations and a transformer manufacturer and the entire United States grid would be down for at least 18 months, probably longer.

Former FERC Chairman Jon Wellinghoff told an audience in 2013: "It wouldn't take that much to take the bulk of the power system down." Fortunately, he went on to explain that rooftop solar and other types of local energy are effective ways to mitigate the inherent vulnerabilities of the Big Grid. He said, "A more distributed system is much more resilient. Millions of distributed generators can't be taken down at once."[35]

Early local energy systems like microgrids were built almost entirely for resiliency—ensuring that electricity continued flowing to critical facilities even when the Big Grid is down. But as costs have continued to plummet, local energy solutions are becoming cheaper than electricity from the grid, and they are now one of the fastest paths to a clean and *reliable* energy future.

770 MILLION PEOPLE WITHOUT ELECTRICITY (#3)

It is still hard for me to believe the statistics—770 million people, more than one-tenth the world's population, have *no access to electricity*.[36] None. This is more than twice the population of the United States living in darkness. About 580 million of those people live in Africa, with another 100 million in India. In the US, tens of thousands of Native Americans live without electricity.[37] This large-scale absence of electricity is mostly due to a lack of grids, but even a grid connection is not always enough to ensure access to electricity. For hundreds of millions more people, electricity is either unaffordable or the service is unreliable. Many areas in India, for instance, electrify their grids for only part of the day.[38] Collectively, these people suffer from what is known as *energy poverty*. The Big Grid has failed all of them.

A few years after my conversation with the Samburu mother, I returned to Africa to see the challenges and solutions firsthand. I went to meet with government leaders, businesspeople, and entrepreneurs to learn how electricity was serving—and underserving—some of the lowest income people in the world. It was an eye-opening trip, and I will share several of the stories from my travels throughout the book.

Most of what I learned was inspiring. Entrepreneurs and innovators are using state-of-the-art technology to create small local energy systems. These are making electricity—and its benefits—affordable to tens of millions of people. The first-generation products, like handheld solar-powered LED lanterns, were modest. But even the ability to safely light a home after nightfall was a major advance.

As the cost of the components declined, the systems grew to include roof-mounted solar panels, bigger batteries, and multiple lights (called *solar home systems,* or *SHS*). Components continued getting cheaper and these off-grid systems added televisions and radios. Solar-powered water pumps and irrigation systems became affordable. The most advanced systems can now support small refrigerators and can even connect neighbors to aggregate the power, increasing and balancing the system's total capacity for all the users.

Several of the families I visited shared the impact electricity had on their lives. Women and children spent less time gathering wood and dung to light their homes. Some families had used expensive and dangerous kerosene lamps for illumination, but electric lights made it practical and affordable for their children to do homework in the evening. Mobile phones could be charged at home, saving many from walking miles to charging stations. One woman told me how the bright steady electric light allowed her to open a small shop that her neighbors would come visit when the sun went down.

The Big Grid casts a shadow over local energy innovations

On that trip I had the chance to meet with the CEO of an African electric utility. I asked him, "How much does it cost you to connect a rural home to your grid?" He answered, "About $1,200." I followed with "And how long will it take to connect everyone in your country to the grid?" He sighed and answered, "We estimate it'll take another ten years, assuming we can get outside financing to pay the upfront costs."

Thinking about all the local energy systems I had already seen, I asked, "How much is it to buy a full off-grid solar home system with a TV and several lights?" Knowing where I was taking this, his shoulders dropped as he answered, "$250." But then he became animated. "There are more forces at work here. We are not a wealthy country, and we rely on outside money in the form of aid and loans to finance electrification. Nearly all that money is specifically tagged for traditional large-scale power plants and power lines."

"Well, haven't wireless phones proven that pulling wires is an outdated requirement?" I asked. I was referring to the fact that mobile phones had allowed most people in Africa to access telephone and internet communications without an expensive, wired infrastructure. He answered, "You might think so, but no. These banks and aid groups are risk averse and very slow to embrace anything that hasn't proven itself over decades." The conversation reminded me that the grid is so big that even its shadow is enough to obscure better solutions.

Fortunately, even in the short time since that trip, the costs of solar systems have continued their decline, and this is forcing perspectives to change. The International Energy Association predicts that local energy solutions like solar home systems and microgrids will be the primary source of power for 450 million Africans by 2030, about one-third of the population.[39]

Making local energy affordable and accessible to everyone

Technology is no longer the limiting factor for local energy. Today, the biggest hurdle is financing. Like cars and TVs, these systems are assets, and they must be paid for before they can be used. In the early days of local energy, these upfront costs were often unaffordable for families on tight budgets. Fortunately, the local energy industry has proven itself to be just as innovative in financing as it is in technology.

For millions of families in Africa, even the $250 price for a solar home system is beyond their means. But innovative financing models, called *pay as you go* (PAYG), have allowed customers to purchase their solar home systems in small chunks, usually one day at a time. Banking via mobile phones is incredibly popular in Africa. Using their phones, people send a small payment, often less than 50 cents (US), each day they want to use their system. Their provider's servers respond with a number code. By entering that code onto a small keypad on the SHS, the battery is unlocked, and they are rewarded with another 24 hours of light and TV. Many of the PAYG products allow people to pay off their system and own it outright after two to three years, after which, the light and TV are free to use.

For the higher income parts of the world like the US, local energy systems need to be much larger so they can power appliances like refrigerators and air conditioners. These larger rooftop solar systems can easily cost $10,000 or more. Even though, over time, they are cheaper than the Big Grid, these upfront price tags are prohibitive for many families. Fortunately, a range of innovative financial solutions have emerged in the US as well. Families with access to credit can finance their systems as they would finance a car. A surprising range of available options—like leases and third-party ownership—can fit just about anyone's needs.

But local energy can still be out of reach for many families. Some do not have the credit. Others live in apartments or rent their homes and are unable to install panels on the roofs. Community solar was created to address both these challenges. A larger solar project is built in the community that can simultaneously serve many homes at once. The business models vary, but people can purchase any number of individual panels, or even rent them. The output of each person's panels is added to the grid, and each family's electric bill is reduced by the proportionate amount. As I will present in Chapter 9, "Unlocking Our Power," community solar is one of the fastest growing types of local energy, and it makes the benefits available to virtually everyone at whatever budget they can afford.

One particularly inspiring company is GRID Alternatives. Its people are laser focused on bringing the benefits of local energy to low-income families across the US. Its mission ". . . is to build community-powered solutions to advance economic and environmental justice through renewable energy." It helps create local jobs through extensive training on solar installation. And it helps design, finance, and install solar for individual homes, multi-family homes, and community-wide systems. As of early 2021, it has installed over 15,000 systems in low-income communities, creating lifetime savings for these families' electric bills of almost $500 million![40]

LOSING STEAM (#4)

Many of the assumptions and constraints that led Edison and Insull to the design of the Big Grid are no longer necessary—or even relevant. Yet the grid we use today remains frustratingly stuck in time. In his book *Energy and Civilization*, Vaclav Smil puts this in perspective: "After more than 120 years the dominant constituents of our pervasive electrical systems—steam turbogenerators, transformers, and high-voltage alternating current (AC) transmission—have grown in efficiencies, capacities, and reliabilities, but their fundamental design and properties remain the same, and their originators would easily recognize the latest variations on the themes they created."

From Thomas Edison's 100 kilowatt plant at Pearl Street Station to a modern 1,000 megawatt nuclear behemoth that is 10,000 times larger, the electricity industry is defined by its scale. But "big" has reached its limits. The age of giant fuel-powered plants is waning. In his book *Smart Power*, Peter Fox-Penner says that the average size of new power plants peaked at 493 MW in 2003 and has been shrinking since.

Why is bigger no longer better? At a certain point, the project management costs for larger and larger plants outweigh the benefits of size, and the added complexity of building a large power plant overtakes the economies of scale. There is no better example of these trade-offs than nuclear power plants, which are among the most complex things ever built.

Nuclear plants epitomize the challenges of big

The engineer in me is inspired by the science of nuclear power. It tames Mother Nature's most fundamental energies. And yet, the businessperson in me does not like them at all. Nuclear plants are expensive, unpopular, and prone to years of delays and huge cost overruns. The reality of building and utilizing nuclear power in the United States in recent years has only supported my skepticism.

Take the unfinished V.C. Summer nuclear plant in South Carolina, which was cancelled in 2017. Already behind schedule, it was forecast to take at least four more years to complete and cost more than double its original $11 billion budget. Its owners wrote off $9 billion—about the same as Rhode Island's entire state budget. And because of laws that protect utility profits, customers of the utilities that made this bad investment will end up paying for much of the money already spent, even though they will not receive a single kilowatt hour from it.[41] Lawsuits are raging, and the CEO of the now-defunct utility that owned the plant will be serving a two-year jail sentence for fraud.[42]

Georgia is home to another struggling nuclear power plant. As of mid-2021, the final chapter of this story remains unwritten, although the narrative already involves staggering cost overruns. Plant Vogtle, the last nuclear plant project in the US, was originally budgeted at $14 billion when construction began in 2013. The project has received $12 billion in federal loan guarantees. That's 22 times more money than the controversial failed solar company Solyndra received from the same loan program, although you rarely hear about the "Plant Vogtle scandal." (More on this later.) Initially targeted for operations in 2017, the Vogtle project has faced a cascade of delays and overruns, with the most optimistic timeline of starting operations now targeted for 2022 with a revised budget of $25 billion.[43]

A 2001 article in *The Economist* titled "A new dawn for nuclear power?" cleverly echoes these issues: "Nuclear power, which early advocates thought would be 'too cheap to meter', is more likely to be remembered as too costly to matter."[44]

Lessons about "big" from the computer industry

Big, as we can see, is not always better. Decades ago, the computer industry faced a similar reckoning from a similar structural problem. It is easy to forget now, but bigger was once better in computing, too. Up through the 1970s, mainframe computers had been the backbone of business computing. They were extremely expensive and required experts to program and operate them. And though the components in them were growing cheaper, their growing complexity ended decades of predictable cost declines (per unit of computing). The personal computer industry was born in this era, and it went on to utterly disrupt the mainframe business and remake the world along the way.

Computing and electric grids have more differences than similarities, but there are still some insights to be gained from comparing them, as I realized during a conversation with a utility executive while researching this book. He was trying to explain the value of the current utility model over local energy to me in terms he thought I might understand. "For many years people owned their computers and ran their applications on them, right?" he said. "But when the internet made cloud-based computing easily accessible, pretty much everyone switched to using applications in the cloud." He was telling me we still need big companies to run things for us. This type of service model, he explained, is basically what utilities have always done.

I quickly realized that he knew less about the computer industry than I did about utilities. Though many forget this, the computer industry originally operated as a "cloud," where people remotely logged into a large central mainframe computer (called *time sharing*). With only a few large companies manufacturing these large machines, the industry struggled to innovate. The personal computer revolution was born in this shadow. Hundreds of companies began making smaller computers, many in their garages. Innovation took off. New software was developed. New uses for computing were discovered. And somewhat ironically, the industry has come full circle toward a cloud model. But this time, the cloud computing services from companies like Google and Apple are made of millions of PC-class machines. Mainframes have practically vanished.

The crucial lesson is this: during the journey from mainframe to cloud that wound through the garages of thousands of innovators, the cost of computing fell an astonishing 100 million times.[45] The same spirit of entrepreneurship and innovation that revolutionized the stagnating computer industry are now coming to bear on the electricity industry. While the cost declines will not be as dramatic, local energy will help drive down costs far faster than the Big Grid could ever accomplish on its own.

Big is a lot less important than it used to be

I talked with numerous utility executives and engineers while researching this book, and many were skeptical of the ability of local energy systems to be a meaningful part of our electricity system. They offered abundant reasons why small systems would always be a niche in a grid defined by big systems. If local energy is to become a core part of our electric future, it will need to address some of perceived benefits of scale in the Big Grid. Let me step through three of the most common arguments in favor of big grids and explain why having a large, centralized system is no longer as important as it once was.

Argument 1: "Bigger grids are easier to balance." Because the options for storing electricity were limited when the Big Grid was created, it is designed to keep the consumption and production of electrons in perfect balance at all times. Every time an air conditioner starts, a tiny amount of electricity must be added to the grid. If supply and demand become too unbalanced, grid-scale circuit breakers will trip, causing brownouts or blackouts. Maintaining that balance is a delicate yet herculean task.

Edison's design accomplished this through sheer scale—any single change in generation or load is dwarfed by the size of the grid. It is like pouring a glass of water into a swimming pool. The only ripples are tiny, and no water will slosh over the pool's side. Edison's design also had the virtue of simplicity, since a few large plants reduce the number of inputs into the system; if you are spinning plates, the fewer the better.

Many skeptics of local energy fear connecting many small independent systems to the larger grid will upset this delicate balancing act. Fortunately, technology has come a long way in the last century. Since every type of local energy contains a smart control system, it can readily be programmed to minimize disrupting the Big Grid, and as I will discuss later, even help improve it.

Argument 2: "Bigger plants are more efficient." Another argument favored by fans of the Big Grid is the supposed efficiency of large power plants. Intuitively this makes sense, but it neglects an important detail: big power plants are remarkably inefficient already. This is the nature of all *thermal power plants,* which use a heat source to create steam and spin an electric generator. Nearly all big plants are thermal, including natural gas, nuclear, geothermal, biomass, and coal plants. For example, even the best natural gas plants throw away more than 30% of their energy as waste heat.

This is not a design flaw; it is real-world limitations imposed by the laws of thermodynamics. Coal and nuclear plants are even worse, losing 60%–70% of the energy they generate as heat, with only 30%–40% being converted to electricity. With fuel being one of the largest costs to a utility, these losses drive up the costs of all thermal plants.

One of the many techniques to improve thermal plant efficiency is to make the plant and its generators larger. While a larger coal or natural gas plant might be relatively more efficient than a smaller coal or natural gas plant, solar does not work this way. Any given solar cell has the same efficiency whether it is powering a small lantern or sitting among millions of others powering a city. This means a solar-based system does not need to be centralized in one large power plant—it can be distributed throughout a system, with appropriate amounts of generation placed wherever they are needed.

Argument 3: "Bigger plants have better economies of scale." The more kilowatt hours a plant generates, the more you can spread out fixed costs, including engineering, design, and permitting. As I said earlier, the principle of economies of scale hits a ceiling at a certain size. As I will share in the next chapter, a new principle is overtaking the power industry. I call it *economies of volume.* It applies to technology-based solutions like solar and batteries, and it is changing the economics of electricity.

SHORT-CIRCUITING OUR CHOICES (#5)

The rules that made the Big Grid possible in the 20th century are now snuffing out the innovations needed to bring it into the 21st century.

Glacial innovation and byzantine policies leave us powerless

I sometimes imagine how much fun it would be to have that time-traveling telephone booth from the movie *Bill and Ted's Excellent Adventure.* It may seem silly, but I would love to meet history's greatest inventors and show them what we have created in the 21st century. I can only imagine how proud Alexander Graham Bell would be seeing an iPhone and how it can instantaneously connect people on opposite sides of the planet and allow them to access the sum of all human knowledge with a few taps of a finger. How excited would Orville and Wilbur Wright be to see a Boeing 747? I am not, however, sure that Edison and Tesla would be as impressed with our society's progress with electricity. If they toured a modern electric substation or visited a coal or hydropower plant, I fear they might exclaim, "It has been 100 years. Is this all the progress you people have made?!?!"

As I further explain in Chapter 7, "Utilities vs the Future," innovation in the power industry has stagnated. With research budgets

1930s

TODAY

FIGURE 2.4 *Some parts of today's Big Grid look remarkably similar to the systems built almost a century ago. Credits: De Luan/Alamy Stock Photo, ikindi/ Shutterstock.com*

at just 0.1% of revenue, utilities spend less on innovation than any other industry in the US. More aggravating, the industry's spending on federal lobbying ranks third among all US industries. Since most utility policy is set at the state level, and most state lobbying is undisclosed, it is a good bet that utilities have the largest lobbying budgets of any industry. In most cases, utilities have no real competition. Because regulators set utilities' profits based on the infrastructure they own, they are guaranteed to have healthy profits as long as they keep building power plants and power lines. In her book, Gretchen Bakke says, "The electricity business is the only one in which you can make a profit by redecorating your office."

One of the most dispiriting parts of my journey into clean energy was learning just how tangled, bureaucratic, and stultifying the policies governing the Big Grid have become. I will offer a brief overview here to illustrate the slog required to modernize US electric policy.

The federal government sets some baselines with key federal laws like the Federal Power Act or the Public Utilities Regulatory Policy Act of 1978, commonly known as PURPA. The feds also have policy groups like FERC and NERC that intertwine with the states. There are regional grid operators setting policy across groups of states with names like PJM, MISO, ERCOT and more. But most importantly, each state has its own unique laws and regulatory frameworks managed by utility commissions. This is where the most action takes place. All these groups create a web of policies aimed at ensuring electric utilities do not abuse their monopolies. In many ways, these policymakers end up locking the monopolies in tighter.

Of these monopoly utilities, there are somewhere between 2,000 and 3,000 in the US, depending on who you ask. Generally speaking, 180 are large investor-owned utilities (IOUs) like Duke Energy and Commonwealth Edison. Another two to three thousand are smaller utilities run by municipalities (munis), public utility districts (PUDs), and cooperatives (co-ops or EMCs). The federal government even owns a few utility-like agencies, such as the Tennessee Valley Authority (TVA), that supply power to other utilities. Many have a monopoly in their regions, and all have their own unique policies, pricing plans, service levels, and mix of power plants.

It requires years of experience to make sense of this industry and decades of experience to attempt the brain surgery of policy change. "The economic and regulatory structure of the American power industry is a contraption only a lawyer could love," quips Peter Fox-Penner in *Smart Power.*

Local energy can restore our power

As individuals, we are virtually powerless to influence the way electricity is generated and delivered to our homes and communities. This point is punctuated by the term that utilities and regulators use to describe their customers like you and me. They call us *ratepayers.* This aptly describes the role we play in their world.

In every other part of our lives, we can sell anything to anyone. We have bake sales, flea markets, garage sales, Craig's List, Girl Scout cookies, eBay, Etsy, even lemonade stands. But when it comes to electricity, it is illegal for anyone other than your utility or some other government sanctioned reseller to sell you electricity. Sure, there are a few pilot programs testing *peer-to-peer electricity trading* in the US, but it is likely more people have climbed Mount Everest than have participated in these trials.

Fortunately, while we wait on the necessary but painfully slow policy changes, there is a way to leapfrog all this. In fact, millions of people are embracing it already. When Insull pioneered and popularized the idea of regulated monopolies, he was focused on ensuring that only utilities could *sell* electricity. He never contemplated that it would one day be practical for people to generate their own power. Local energy is rising, and it is doing so in the ways that the Big Grid planners never knew could exist.

But is local energy ready to compete with the reliability and the cost of the Big Grid? As utilities and regulators attempt to slow local energy with fees and restrictions, can local energy offer enough benefits to overcome low public awareness and the intransigence of the incumbent power industry? I will answer these questions and more in the next chapter.

CHAPTER 3

THE RISE OF LOCAL ENERGY

It was October 8, 2017, and one of California's large seasonal wild-fires was edging closer to the Stone Edge Farm and Winery. As they were for hundreds of other vineyards in the Sonoma region, the wildfires were an existential threat to Stone Edge, poised to destroy crops Stone Edge had spent years sustainably cultivating. The fire was also causing power outages across the region, which shut down essential agricultural systems. Even if the fire spared it, should the farm lose power, the irrigation pumps would shut down, putting the crops at risk.

Stone Edge, however, was ready. At 5:00 a.m. on October 9, the team overseeing the farm decided this was the moment to test the full capabilities of an electrical system they had spent years develop-ing. They disconnected the farm from Pacific Gas & Electric's grid and put the farm's microgrid into *island* mode, letting the solar and battery systems generate 100% of the farm's electricity.[1] The farm was able to keep the pumps and other core systems running, entirely independent of whether the Big Grid was working. The system allowed the microgrid operators to control every aspect of the system safely, miles from the farm and the approaching fire. For another ten days, until the threat of the wildfire had passed, the Stone Edge microgrid powered every aspect of the farm.

It was an amazing story of innovation and resilience in the face of powerful natural forces and utility shortcomings, and it earned the system's designers accolades from the industry (and helped save the crops, too).

A few years after the 2017 fire, I traveled to Stone Edge Farm to see this famous microgrid firsthand. I had not expected that the beauty of the farm was worth the trip on its own. From its cobblestone roadways, manicured plants, gorgeous buildings, and even an iconic two-story observatory and telescope, it is naturally beautiful, agriculturally functional, and technically state-of-the-art. Spread across the 16-acre property were various components of the farm's microgrid. These batteries, solar panels, and other components of the microgrid were blended into the farm so well that I had to go out of my way to see them. Not only was the Stone Edge microgrid one of the most advanced systems of its type in the world, it was also one of the most beautiful.

The idea for this groundbreaking system began taking shape in 2013. The farm's owners, retired financial executive Mac McQuown and his wife, Leslie, wanted to make their small farm even more sustainable by powering it with carbon-free, renewably generated electricity. But rather than just adding solar panels, Mac and Leslie decided to pioneer a next-generation local energy system.

The Stone Edge microgrid started where most such projects start—with solar panels on a rooftop. Now, there are 12 solar arrays on 10 buildings for a total of 550 panels generating 160 kilowatts, enough to power 40 homes.[2][3] The microgrid employs multiple types of batteries, including five of Tesla's grid-scale Powerpacks, traditional lithium-ion batteries, and specialized saltwater batteries from a company called Aquion.

One particularly cutting-edge part of this microgrid is the "Hydrogen Park," which uses excess solar power to break apart water molecules ($H2O$) through a process called *electrolysis*. The freed oxygen is released into the air, but the hydrogen is captured, stored in tanks, and used to fuel specialized hydrogen-powered automobiles from Toyota and Honda, giving the farm clean mobility powered by sunlight. Excess hydrogen is stored in high-pressure tanks so it can later be turned back into electricity using a technology called a *fuel cell*. The round-trip conversion of electricity to hydrogen and back again is effectively a long-term battery that augments the traditional batteries and the solar panels during extended cloudy weather. (I will look at these technologies later in the book.)

The resulting microgrid, an engineering marvel, has become a proving ground for several energy tech startups and a living laboratory for dozens of universities whose students come to study and help optimize the system. But there is one more technology that makes this microgrid truly unique.

Like the Big Grid, most microgrids are managed centrally, their various components subject to the directions of a single control system. But technology from a company called Heila takes a very different approach, distributing decision making throughout the system. Each battery, each solar panel and inverter, and each part of the Hydrogen Park is its own responsive, independent node in an interconnected network of peers. Using what mathematicians call *game theory*, each node watches the others, and they coordinate in real time to optimize their resources to maximize the effectiveness of the entire system. If you come from the computer industry, a system design like this is commonplace. But for electric systems, where resistance to change is pervasive, this is a radical new idea.

Heila's unique distributed decision-making architecture has given the Stone Edge microgrid unprecedented flexibility and resiliency. For instance, when a new component, like a battery, is added to a traditional microgrid, the entire system needs to be redesigned to ensure it remains balanced and safe. When a new node is added in Heila's technology, the other nodes sense it and dynamically balance their own resources, keeping the overall microgrid optimized. This architecture is extremely resilient. If any component or node fails, the other nodes just re-balance to compensate, avoiding a system-wide shut down. This is more than a radical new approach to microgrids. Heila and Stone Edge are proving that electricity can be managed in a far smarter, more reliable way that will almost certainly influence the Big Grid in years to come.

Mac told me, "Our marginal cost of electricity is zero. Our marginal carbon footprint is also zero. Our electric bill is zero. The only waste we create is a little heat. The entire economics are in the front-end capital investment and the cost to finance it. And these costs are coming down quickly. There is no other way to get energy at this cost and with zero waste. I believe every building could, and should, be its own microgrid. The current electrical power industry is just an artifact of history."

Mac and Leslie are backing up this vision with their time and money. Because their microgrid is the first of its kind, everything is expensive. They are taking risks, often becoming the first customers of energy tech startups like Heila, and they are taking what they have learned and sharing it with hundreds of scientists, startups, and policymakers. Each of the early components they incorporate represents the potential for tens of billions of dollars of opportunities as the technologies become mainstream. It seems fitting that the Stone Edge microgrid is just a few hours' drive from the garage where Steve Jobs and Steve Wozniak created the first Apple computer.

In August 2018, Leslie pressed a button and set the microgrid into permanent island mode, creating an indelible chapter in the history of local energy.[4]

LOCAL ENERGY 101

You can tell a lot about how different people think about an idea by the language they use to describe it.

Engineers, in their typical matter-of-fact way, refer to local energy as *direct-to-load*, and *distributed energy resources* (DER), because relatively tiny power plants, like solar panels on a roof, are scattered throughout the grid, each directly powering a home or building.

Policymakers call it non-wires alternatives (NWA), grid edge, distributed generation (DG), and behind-the-meter.

Renewable energy advocates emphasize that it is *democratized energy* or *self-generation*.

And utility executives have their own colorful language that includes scary terms like *grid defection* and *death spiral*, along with some more familiar but otherwise unprintable names.

Each of these terms is rooted in the interests of the groups that design, build, and maintain our power systems. But there is one group that has no voice and no words for this—the people and businesses that consume the electricity—the customers. Throughout the book, *local energy* is not just the way electricity is generated, it also reflects the benefits to the people and communities it serves.

The components of a microgrid

Before solar became so cost competitive and local energy was more expensive than the Big Grid, early microgrids relied on fossil fuels. One particularly innovative approach still widely used today is called *combined heat and power* (CHP). These fossil-fuel powered microgrids generate electricity and useful heat at the same time. In some cases, the heat is already being generated, typically from natural gas, for an industrial process like chemical manufacturing. In other cases, powering an electric generator is the primary use for the heat. Either way, the generators throw off their own excess heat, which is then used to heat a building. Because these systems combine so many benefits, including powering a building when the grid is down, they are more efficient and often less expensive than paying for separate systems.

More recently, solar and batteries have become the most common generation method for new microgrids. In addition to being cleaner, these systems require no fuel and can run indefinitely during extended grid outages. The most common form of these microgrids is smaller *residential solar and battery systems.*

Larger, commercial-scale systems, like the Pedro Fernandez microgrid in Puerto Rico are gaining traction. Very large-scale systems are making news. The Hawaiian Island of Kauai and the town of Minster, Ohio, are both powered by enormous solar and battery local energy systems.

All these systems share a similar design, including four primary components: solar panels, batteries, a controller, and a switch that can disconnect the home or building from the grid (see Figure 3.1).

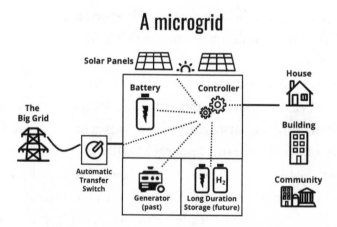

A microgrid

FIGURE 3.1 *A simplified solar microgrid is an integration of several components. When "islanded," it can operate independently of the Big Grid and provide some or all the power for a building or home.*

The central heart of a microgrid is its controller. This is like a symphony conductor coordinating different instruments. An automatic transfer switch sits in between the Big Grid and the microgrid. Disconnecting the switch puts the microgrid in *island* mode, allowing it to run independently of the grid. During the day, the solar panels power the building and charge the battery. At night, the battery discharges, providing uninterrupted power. While early generations of microgrids relied on fossil fuel generators, future microgrids will use long-duration storage, which I explore in Chapter 5, "Hidden Patterns of Innovation."

Most microgrids have capacity limitations, so they cannot power every load in a home or building at once. For example, the Pedro Fernandez microgrid is not large enough to power the entire school. The designers focused on the subset of electric loads necessary for the school to remain open, called *critical loads*. One part of the microgrid directly powers the water and sewage pumps, ensuring the school remains habitable. Another part powers the lights, computers, telephones, and networks, and a big refrigerator in the cafeteria. As I walked through the school back in 2019, I noticed a few of the electric outlets were bright red. I learned that these outlets were directly connected to the microgrid. When the grid goes down, these outlets remain energized, providing 24/7/365 power for anything plugged into them.

Challenges remain

Why are local energy systems like the one at the Pedro Fernandez school the exception today rather than the rule? Some of the challenges are straightforward. Not all roofs are strong enough to host solar panels. A more common challenge is cost. Systems with batteries can still be more expensive than electricity from the Big Grid. Even when these systems are cheaper, many organizations and families lack the credit to finance such a system.

As I will show throughout the rest of this chapter, the costs of these systems are rapidly declining, making them increasingly affordable to everyone. Three years ago, the Pedro Fernandez microgrid would have been more expensive than grid power. In 2020, it was roughly the same price (referred to as *grid parity*). In a few years, a new microgrid will be cheaper than the grid.

Think about flat screen TVs, which followed technology cost curves similar to the ones solar and batteries are on now. In 1999, first generation flatscreens cost $25,000. But as flat screen prices declined year after year, they came to dominate the market and cathode ray tube (CRT) televisions faded into history. Today, a high-quality flat screen TV often costs less than $500.

Local energy pioneers also must contend with the technical complexity of designing first-generation systems. Fortunately, local energy systems are evolving quickly. Standards are emerging. These systems are well on their way to being just as manageable to install and own as your air conditioner. As I share below, the *consumerization* of local energy is making local energy accessible, affordable, and easy enough for everyone to use.

Solar panels without batteries fall short of local energy's greatest promise. Of course, they are great for lowering electric bills and helping the environment. But batteries are needed for solar to island and operate without the grid. Across the world, electric safety regulations require panels to be automatically shut down during power outages, as many Puerto Ricans with rooftop solar learned after Maria. This is done to prevent back-feeding electricity onto the grid and injuring the utility workers attempting to repair damaged power lines.

The biggest challenge of all is that local energy is a new idea. Many people struggle to imagine electricity beyond the Big Grid. It is easier to fall back on outdated answers than to embrace new paradigms. Often these perceptions are genuine misunderstandings.

But, as I will share in Chapter 8, "The Battle for Public Opinion," some of these misperceptions are intentionally propagated by an industry that is working hard to maintain monopoly control.

The dollars and cents of local energy

You probably do not think about your electricity bill too much. You are what your utility calls a *ratepayer*. You have almost no choice about what you pay, but you can take comfort that the rates were negotiated by your utility and a government body, usually called a *public service commission*. If they have done their jobs, your electricity costs are fair and equitable.

But with local energy systems, you get options. Chief among your newfound choices is how to build and pay for the local energy system.

The time it takes to pay back the construction costs of a local energy system is driven by several factors, including local policies, the actual cost of installation, and, of particular importance, the regional price of grid electricity.

You might be surprised to learn just how much the price of electricity varies across the US. Hawaii has the most expensive electricity at 28.7 cents per kilowatt hour (the state must ship in fuel, primarily oil, to feed its power plants). The average price in the US in 2019 was 10.5 cents, with California at 16.9 cents and Louisiana with the cheapest, at 7.7 cents.[5]

When the Pedro Fernandez microgrid was designed, the cost of residential electricity in Puerto Rico was 20.3 cents per kilowatt hour. With Puerto Rico's median household income being just one-third that of the rest of the US, the burden of electricity on Puerto Ricans' monthly budgets is significantly greater than that of the rest of the US. For commercial buildings like schools, the price of electricity in Puerto Rico was even higher, at 22.6 cents per kilowatt hour, twice the US average.[6]

Most local energy systems are financed and paid for over time. The effective cost per kilowatt hour is calculated by dividing the annual financing payments by the number of kilowatt hours the system generates that year. The Pedro Fernandez microgrid is illustrative. Although most of the cost was donated, had it been financed like most systems, the school's microgrid would work out to be 20 cents per kilowatt hour over a 20-year payback—expensive, but still a bit cheaper than what the school pays for its grid electricity. And as Puerto Rico's cost of electricity inevitably rises (the island's utility is currently recovering from its 2017 bankruptcy), the fixed costs of the microgrid will be increasingly cheaper by comparison.

The best source of data on solar costs, financing, and payback time is a site called EnergySage.com. I asked CEO Vikram Aggarwal about the basic economics.[7] He told me, "The typical payback time for rooftop solar in the US is 5 to 10 years, depending on where you live." But he added, "Since solar panels often last much longer, sometimes more than 30 years, the electricity generated after the payback period is free." Free is the best price of all, and the story on costs is getting even better.

What makes up the costs in a local energy system?

The full price of building a system like the Stone Edge or Pedro Fernandez microgrids can be broken down into three large cost buckets. The first is *hardware*, like solar panels, batteries, and inverters. These have all experienced incredible price declines and are one of the biggest reasons local energy has reached grid parity in recent years. The second cost bucket is labor. Here again, a steady flow of new products and techniques are making the job of installing local energy systems more efficient, safer, and lower cost. The final bucket of cost, called *soft costs*, includes a list of administrative items like interconnecting to the grid, financing costs, system design, project management, permitting, sales tax, and installer profits.

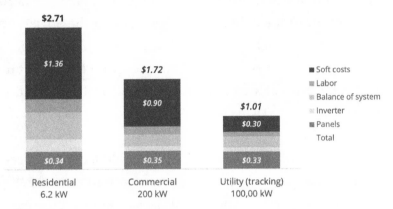

Soft costs are the largest part of small US solar installations
(US$ costs per watt in 2020)

FIGURE 3.2 *Soft costs make up the largest single part of local energy installations. Source: National Renewable Energy Laboratory (NREL) (freeingenergy .com/g133).*

The National Renewable Energy Laboratory (NREL) compared the cost breakdown across residential, commercial, and utility-scale systems (see Figure 3.2). It found that soft costs make up just 26% of utility-scale projects but are roughly 50% of the cost of residential and commercial-scale systems. (This data is for solar-only systems; batteries add about $1 per watt for mid-scale systems.)

It is hard to believe that half the costs of a system like the Pedro Fernandez microgrid have nothing to do with generating electricity. Even more frustrating, these high soft costs are unique to the US. Other countries have cut out the bureaucracy of local energy soft costs and are building identical systems for 20% to 50% less than the US.

Soft costs have remained stubbornly fixed over the years, but this is finally changing. As we will see in Chapter 7, "Utilities vs the Future," the local energy industry is stepping up and finding creative ways to drive down these costs.

SMALL IS THE NEW BIG

In the age of digital technology and low-cost renewables, the rationale for a grid built exclusively on economies of scale is outdated. The future of the grid will be built with a mix of technologies, owners, and business models—at every scale, from small to large. While microgrids and other forms of local energy do not benefit directly from economies of scale, the list below offers 18 new and unique benefits that more than compensate.

1. *Increased reliability.* As I said earlier, 90% of all outages come from the power lines that connect you to centralized power plants. When the "power plant" is on your roof, this single point of failure is removed. Many local energy solar projects use a technology called *microinverters*. They remove virtually every single point of failure in these small systems.

2. *Lower costs.* Thanks to decades of price declines in the solar panels, batteries, and electronics, it is increasingly cheaper to generate electricity onsite than it is to buy it from the grid. In the next chapter, I explain how component costs will continue declining for decades to come, ultimately making local energy systems like microgrids the cheapest source of electricity for everyone, everywhere.

3. *Faster to build.* Local energy systems can be built in weeks and even days. Construction times for large power plants, on the other hand, are measured in years. Nuclear plants take an average of 9.8 years.[8] In 2020, the US installed enough residential

and commercial solar to match the energy output of a full nuclear plant.[9]

4. *Easier to finance.* There are few institutions that can finance billion-dollar power plants and transmission projects. But financing for small-scale systems, many of which cost less than a compact car, is widely available from virtually every kind of lender. As renewable energy entrepreneur, investor, and now Director of the Department of Energy's Loan Program, Jigar Shah, writes in his 2013 book, *Creating Climate Wealth*, "It's easier and more efficient to deploy 1,000 $1-million projects than deploying a single $1-billion project. . . . Small projects are easier to get off the ground. They're not bogged down by bureaucratic red tape and endless approval layers. Small projects don't require a multi-stakeholder process."[10]

5. *Creates local jobs.* Installing local energy projects like rooftop solar and microgrids creates good-paying jobs that cannot be outsourced. Across the US, clean energy jobs like these pay almost 30% higher than the national median hourly wage.[11] While utility-scale projects represent a larger portion of solar capacity built each year, 80% of the jobs in solar installations are actually for smaller scale, local energy projects.[12] Per megawatt of installed solar, local energy projects create 11 times more jobs than utility-scale installations.[13] The US Bureau of Labor projects solar installers will be the third fastest growing job in the coming decade, with an average pay of $45,000 per year.[14] Most of these jobs will be in the communities where the installers live. It is like farm-to-table for electrons.

6. *Benefits communities.* The paychecks of the local installers feed a virtuous cycle for the community. Employees buy from local

stores and pay the taxes that support the community's law enforcement officers, firefighters, teachers, parks, and schools. One study found that a quarter of the money spent on residential solar projects stays within the community.[15]

7. *Fosters social equity.* For the nearly one billion low-income families in Africa and India with no access to the Big Grid, local energy is the only way to enjoy the benefits of electrification. It is the first and most practical step-up—what economists call the *energy ladder*. Even within the US, low-income households and communities can benefit from local energy. Rather than siphoning off the profits from electricity to giant corporations and Wall Street, local energy keeps some or all the financial returns within the families and communities that need it most.

8. *Helps the planet.* Many people are looking for ways to reduce their personal carbon footprint. The three biggest categories of household carbon emissions are food (17%), personal automobiles (23%), and electricity (25%).[16] While avoiding meat and taking public transportation make a big difference, it is arguable that rooftop or community solar are the largest and most direct ways for most people to reduce household carbon. (That is, unless you want to ride a bicycle everywhere.)

How much CO_2 and pollution are reduced when solar replaces electricity that would otherwise be created by the fuel-burning power plants of the Big Grid? Even a small system like a residential solar rooftop can have a surprisingly large impact. Assuming an average rooftop generates one megawatt hour of electricity each month and that megawatt hour were to directly offset electricity from a coal plant, it would keep 1,100 pounds of coal in the ground. It would also eliminate 1 ton of CO_2, 185 pounds of ash, 3.9 pounds of sulfur dioxide, 1.6 pounds of nitrogen oxide, 1.7 ounces of radioactive thorium and uranium (yes, coal contains radioactive elements), and 13 mg of particularly nasty mercury. Every month that solar rooftop operates, a coal plant needs 16,000 fewer gallons of water for cooling.[17] (Nuclear plants use a similar amount of water for cooling.[18]) Replacing natural gas saves about three-quarters of a ton of CO_2 and methane (CO_2 equivalent) every month.[19] These are like planet-saving dividends—they pay back these pollution reductions every single month again and again, for as long as the solar installation operates.

9. *Creates choice.* The electric monopolies force all buying and selling through a single organization. In most cases, this results in a marketplace of one. Customers have no choice but to accept whatever rates and service levels their monopolies choose to offer. Conversely, local energy is like buying a car or a home renovation. People can shop for options and prices and choose the one that best suits their interests.

10. *Spurs innovation.* Millions of businesses and individuals are purchasing these local energy systems. Hundreds and thousands of companies, big and small, are competing to offer the best options to this enormous, growing customer base. Competition spurs innovation and creates a virtuous cycle of value to customers.

11. *Isolates risks.* Like the famed garages of Silicon Valley startups, local energy projects offer a sandbox for entrepreneurs and innovators. Innovation flourishes when risks can be taken, especially when the impact of a failure is isolated to a single project.

12. *Increases security.* Cyberattacks on grid infrastructure are on the rise. Widespread attacks are a matter of *when*, not *if*. Fortunately, local energy provides what the tech industry calls *security through diversity*, which means it is much harder to attack a system made of many small diverse parts than one comprised of a few large systems.

13. *Avoids transmission delays and costs.* Adding new utility-scale solar and wind projects to the Big Grid requires a complex, drawn out, expensive approval process. Each new project must be placed in an *interconnection queue* for analysis. A report from Lawrence Berkeley National Labs (LBNL) found that it takes an average of four years from queue request to completed project![20]

    These delays are crushing to developers and investors; 75% of projects drop out and are never built. Additionally, in most cases, the project developers must cover the costs of any new transmission lines required to connect their project into the Big Grid. Another study found this increases total project costs by 50%–100%, undermining many of the cost advantages utility-scale projects have over local energy.[21] Small local energy projects do not use the transmission grid and avoid these problems altogether.

14. *Reduces transmission losses.* Transmission and distribution power lines lose about 5% of all the electricity they deliver due to heat loss. Every kilowatt hour generated locally avoids these losses and goes directly to customers.[22]

15. *Stabilizes the Big Grid.* Solar microgrids are smart. They can not only coordinate with the Big Grid to balance power demand and diversify generation sources, but they are starting to offer a

range of technical services that previously required expensive investments by grid operators. As I will explain later in the book, services like *demand response, frequency regulation,* and *virtual power plants* demonstrate that local energy need not be disruptive to the Big Grid. In fact, the exact opposite is true—when embraced intelligently, local energy can help solve many of the Big Grid's challenges.

16. *Creates transparency.* When the flow of money is controlled by a small set of executives, policymakers, and legislators, shady shortcuts become tempting and easy to hide. Open markets enforce transparency, honesty, and fairness.

17. *Makes batteries more valuable.* Batteries and other types of electric storage are useful anywhere they exist on the grid. Beyond just storing electrons, they can provide many technical *ancillary services* like voltage regulation and frequency regulation. A 2015 paper from RMI makes the case that batteries are *most* valuable when used locally, at a home or a local business, for services like frequency regulation and peak shaving. The paper explains: "Customer-sited, behind-the-meter energy storage can technically provide the largest number of services to the electricity grid at large . . . regulators, utilities, and developers should look as far downstream in the electricity system as possible when examining the economics of energy storage."[23]

18. *Locks in electricity rates.* The price of a kilowatt hour from the Big Grid has been rising for years in the US, even faster than the inflation rate.[24] Solar and battery systems cost almost nothing to operate. Their only costs are payments on the loans used to build them. These finance payments are locked in when the system is built and will not go up over time. Best of all, when local energy systems are paid off, each kilowatt hour generated from then on is effectively free.

Cost comparisons to large-scale solar miss the point

Despite all the advantages of small, distributed systems, it can still be hard to get people to shake Big Grid thinking, even people who are enthusiastic about the solar technologies that make local energy so great. I encountered this resistance in a conversation with a well-intentioned utility executive who, in trying to sell me on the virtues of a large, grid-scale solar project, attempted to give me a lesson in economics. "Listen, Bill," he told me, "If it makes people feel like they are helping with the environment by putting panels on their roofs, then that's fine. But I'm a dollars-and-cents kind of guy. No matter how you pencil it out, if we want to reduce our carbon footprint, then large solar plants will always be the cheapest way to go."

I thought about his answer for a moment. Then I replied, "It may be cheaper for you, but it is not cheaper for me." There was silence. I went on. "The cost of a kilowatt hour from my rooftop solar is already cheaper than what I pay my utility. No matter how much cheaper each kilowatt hour might be if my utility builds its giant solar project, little, if any, of its savings will trickle down and lower my electric bill. If my neighbor puts solar on her roof next week, every kilowatt hour it generates will be cheaper than the rate she pays the utility."

There was an awkward silence. Then he changed the subject.

Variations of this dialog have come up several times as I interviewed people in the industry. People invested in the Big Grid are struggling to see past this simple logic: lower costs for the grid are unlikely to result in lower costs on grid customers' electric bills.

There are several reasons for this. First, the lower cost of utility-scale solar is somewhat offset by the cost to deliver it—the transmission and distribution of a solar kilowatt hour can cost more than generating the solar kilowatt hour in the first place. Second, utilities have huge sunk costs, like existing power plants and transmission upgrades that drown out any savings from a new solar (or wind) plant. Third, these sunk costs do not take into account the hundreds of billions in cleanup costs from 50 years of toxic waste piling up outside their legacy power plants.

The regulators, utility executives, and legislators have been so steeped in Big Grid thinking that they struggle to understand that for the rest of us—homeowners, local businesses, neighborhoods, and communities—building our own local energy systems is *cheaper.*

John Farrell of the Institute for Local Self-Reliance (ILSR) puts this into simple economic terms I really like: "Local energy competes in an entirely different market."

CONSUMERIZATION TURBOCHARGES LOCAL ENERGY

Millions of local energy systems have already been built, and the rate of new installations is increasing. Costs are plummeting for components like solar panels, electronics, and batteries. Standards and plug-and-play products are making it easier to design, purchase, and operate these systems. Together, these trends are leading the local energy market into a new, disruptive era that I call the *consumerization of energy.*

Data from NREL shows the stunning cost declines over the last decade (see Figure 3.3). From 2010 to 2018, the cost of solar installations plummeted between 2.5 times and 4 times!

What is behind these declines? Most prominent are the plummeting costs of solar panels and inverters. Labor costs are also going down thanks to constantly improving techniques, tools, and products.

The declining costs of solar projects
(US$/watt, 2010-2020)

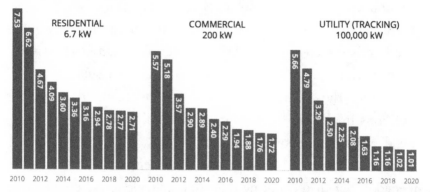

FIGURE 3.3 *Solar project costs, 2010 through 2020. Source: National Renewable Energy Laboratory—Solar Installed System Cost Analysis (freeingenergy .com/g113).*

Let me give you an example. In 2014 I met with the CEO of Zep Solar, which had recently been acquired by Solar City for $158 million. (Solar City, and Zep Solar with it, are now subsidiaries of Tesla.) What did Zep Solar make that was so interesting to Solar City? It invented a mounting system that cut the time it takes to install a solar rooftop from two or three days, down to one day.[25] This not only lowered the labor costs of a solar installation, but it also meant that a single crew could install two to three times more roofs per month. In Chapter 10, "Powered by Innovators," I will introduce many of the innovators that are driving the cost of local energy lower and lower.

There is still a lot of room for cost improvements. For instance, when the RMI team led the design of the Pedro Fernandez microgrid, they faced a quandary countless others have faced before them, one captured by an industry joke: "If you've seen one microgrid, you've seen one microgrid." Nearly every local energy system, both residential and commercial scale, is designed and constructed one at a time. Like a home renovation, customers expect a lot of customization, which means soft costs like engineering design and project management become an outsize part of the overall costs.

This is even more complicated when trying to make different electronic standards and equipment work together. Local energy is still far from achieving the plug-and-play simplicity that will unleash its full potential. True, there are a handful of standards for wiring panels and inverters, but much of the design and installation is different for each project. Newer components like batteries and battery inverters have even fewer standards.

Designing the Pedro Fernandez microgrid was a bit like wiring up a 1990s computer with its crazy range of incompatible ports like RS-232, VGA graphics, and phono jacks for audio. Back in the day, it took expertise and a lot of patience to wire up one of these computers. But after many years, new standards like USB emerged and most of these ports faded into history, along with all the headaches they created. Similar standards are being created for local energy. In a few years, connecting panels, inverters, batteries, and smart home controls will be as easy as USB and Bluetooth.

As we have seen in other technology-driven markets like computers, media, mobile phones, and soon, electric vehicles, consumerization is a tipping point. It breaks the incumbent's lock on markets and unleashes waves of innovation. Garage startups become billion-dollar companies. Tens of thousands of entrepreneurs and investors flock to an explosive set of opportunities for hundreds of millions of brand-new customers. Almost overnight (or so it seems), individual consumers gain access to technology that was once the sole domain of large corporations, governments, and elite experts.

And this consumerization process will pave the way for decades of cost declines in local energy. Here is a short list of emerging trends that will make local energy low-cost and mainstream.

- *Plummeting component costs.* Solar panels, batteries, and electronics are on a precipitous price decline that is expected to continue for decades. This trend is fueled by years of scientific breakthroughs as well as something called *economies of volume*, which I describe in the next chapter.

- *Pre-building for local energy.* The wiring costs of local energy systems can drop by ten times or more when homes and buildings are designed with local energy in mind. This is one of several ideas I advocate for in Chapter 7, "Utilities vs the Future."

- *Building Integrated Photovoltaics* (BIPV). Today, solar comes in a single form, large rectangular panels that are anchored to your roof. But the next generation of solar is not added to your roof, it *is* your roof (and your windows, walls, driveways, and awnings). I explain BIPV in Chapter 5, "Hidden Patterns of Innovation," and explore the business opportunities in the final chapter.

- *Modular systems.* In 2020, Tesla launched an innovative new pricing approach for rooftop solar panels. Rather than each project being unique and requiring bespoke design, Tesla offered it in three simple sizes. This is one of the early industry shifts from metaphorically molding with clay to building with LEGO bricks.

Consumerization is already transforming the economics of the smaller scale systems used in low-income parts of Africa. Before 2017, nearly all the systems sold were tiny solar-powered lanterns. As the cost of solar, batteries, and electronics plummeted, a new market for *solar home systems* emerged, raising living standards for people throughout the continent. In the three years that followed,

3 million systems capable of powering not only lights, but radios and even televisions were shipped.[26]

Local energy is already cheap. Its costs will decline 3 to 4 times further.
What do all these cost declines mean? How far can they go? According to multiple experts, including NREL, the cost of residential-scale local energy will plummet to as little as one-quarter of today's cost (see Figure 3.4). Equally important, the price of electricity from the Big Grid is expected to rise slightly—so the price advantage of locally generated energy is going to become irresistible.

Price of residential solar vs grid electricity
(cents per kilowatt hour, 2010-2050, in 2019 dollars)

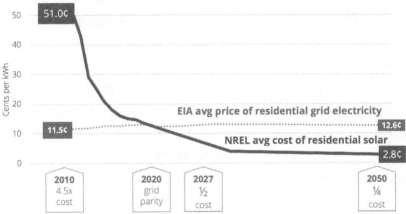

FIGURE 3.4 *The cost of residential solar has declined sharply over time, reaching parity with the average price of grid electricity in 2020. It is projected to reach one-quarter the cost by 2050. Source: NREL ATB 2020 (freeingenergy.com/g114).*

It is hard to overstate how important and disruptive this cost decline will be to the power industry. Saving 5%–10% in 2021 may not be enough to motivate building owners to install local energy. But by 2028, a home with moderate direct sunlight can generate its own electricity for *half of what the owners pay their electric utility!*

(This is based on a US average price of electricity. The gap will be bigger or smaller, depending on each local region's electricity prices.) The urgency to adopt local energy will only increase as grid prices inevitably increase and local energy systems get cheaper.

THE LOCAL ENERGY DECADE

Energy transitions usually take 50 to 75 years. Local energy could be different.

Creating an electricity system that is free of fossil fuels is a formidable goal. In the past, energy transitions of this magnitude have taken decades. The energy historian Vaclav Smil makes this argument in his book *Energy and Civilization*. (Bill Gates lists this book as one of his all-time favorites, and I share the sentiment.) Smil has analyzed energy transitions from burning wood to coal, from gas lighting to electricity, and from coal to nuclear. He concludes that energy transitions take 50 to 75 years, and that achieving a world powered by clean, renewable energy will take decades longer than optimists want to believe.

But things could be different this time. Each of the last few energy transitions required enormous capital, along with the support of giant corporations and governments. In other words, every previous transition was built with economies of scale.

For example, the transition from steam power to electricity required giant corporations to finance mining and drilling, complex distribution networks to deliver the extracted fuels, and huge power plants to wring out every bit of efficiency as fuels were burned. These were intricate systems that required billions in capital and a generation to construct. They required hundreds of laws and financial incentives to weave all the pieces together. In contrast, small local energy systems like solar+battery simply do not require infrastructure and investment at any similar scale. Even setting up a giant solar panel factory is a comparatively simple and affordable enterprise compared with building a nuclear plant or extending transmission power lines across states.

Local energy from wind, water, and earth

Much of the progress of local energy has been driven by the plunging costs of solar, and more recently, batteries. While these will remain the largest source of local energy across the world, there are several other mechanisms to generate electricity vying for market share in the local energy revolution.

Geothermal. This source of energy lies just under our feet. Geothermal power plants were traditionally built near the edges of tectonic plates where high temperatures were close to the surface. But recent advances in drilling technology may expand the locations where these power plants make economic sense. Because these plants are *thermal*, like coal or natural gas, they are usually large and centralized. However, a Swedish company, Climeon, offers a smaller-scale system that generates electricity from the less intense heat available closer to the surface, but its technology has not yet been widely adopted.

The most widespread use of geothermal energy, especially for local-scale systems, is for heating and cooling buildings, rather than generating electricity. Industry leader Dandelion explains that the temperature of the earth 10 feet below the surface is a constant 55°F.[27] This thermal stability is used to dissipate hot summer air and to help warm up cold winter temperatures.

Wave and tide power. There are hundreds of pilots around the world tapping restless ocean energy to generate electricity. These systems are, of course, limited to locations near oceans. And at least so far, their costs are prohibitively high, coming in around three-to-five-times those of solar, wind, and natural gas. Ocean powered local energy is an attractive idea, but it is unlikely to expand beyond a niche solution.

Distributed wind. In the 1920s, a company called Jacobs Wind Electric commercialized small-scale wind turbines and brought electricity to 20,000 customers, primarily rural farms.[28] But these small wind turbines are relatively inefficient compared with their larger, 300-foot brethren. Economies of scale have made wind the largest source of renewable energy in the world (so far), but it also limited the cost savings achievable with smaller-scale wind systems. You may not find many small wind turbines in your neighborhood, but a company called One Energy is bringing the giant wind turbines to larger, industrial customers. Placing a large wind turbine next to a factory, particularly where strong winds are common, can offer cheaper and cleaner electricity than a factory or office park can buy from the grid.

Small modular nuclear reactors. Nuclear plants epitomize the large, centralized nature of the Big Grid. But a new generation of small modular reactors (SMRs) promise to be much safer and a bit cheaper than their behemoth cousins. A company called NuScale Power, the only SMR company approved for testing so far, promises a system that can be built in factories, and like natural gas turbines, shipped to their final site on rail cars. Each unit can power a small town. Nuclear power offers a steady, *dispatchable* source of electricity that is easier to manage than *intermittent* power from solar and wind. But for SMRs to become a mainstream source of local energy, they will have to overcome two challenges.

First, advocates will need to convince an anxious public that it is okay to have a nuclear plant nearby. Second, SMRs are still many years away from wide availability. NuScale's initial pilot is scheduled for the end of the decade. When its reactors are ultimately approved for broad commercial deployment, perhaps another 5 to 10 years after that, they will be competing with far cheaper solar and batteries that together make solar dispatchable, removing one of the SMR's biggest advantages.

Fuel cells. The most common way to convert natural gas or hydrogen to electricity is with a combustion generator. Similar to their diesel generator cousins, the fuel is ignited and gases expand to turn a crank that then turns an electric generator. Fuels cells are a scientific breakthrough. They use a chemical catalyst to directly combine hydrogen and oxygen without combustion, generating far more electricity per unit of fuel than traditional turbines and generators. Current generation fuel cells are too expensive for all but niche applications, but billions of dollars of research are pouring into the market. Experts are optimistic that hydrogen-powered fuel cells will become a mainstream part of the clean energy future, including local energy applications like the one in the Stone Edge microgrid. It is unlikely that hydrogen and fuel cells will replace batteries outright, but they do promise to provide a compelling complement that can store excess solar for days and weeks.

Distributed hydropower. The backbone of the Big Grid is built on enormous hydropower dams like Niagara Falls and more recently, the 22-gigawatt colossus in China, the Three Gorges Dam. Traditionally, hydropower has benefited from economies of scale and has struggled to be cost-effective at smaller scales. But a new breed of small-scale *distributed* or *micro hydro* technologies are promising around-the-clock, dispatchable electricity at prices that approach local solar and batteries. Emily Morris, the CEO of distributed hydro pioneer Emrgy, explains: "It is like a water-based version of two small wind turbines. They can be dropped in place without anchoring and with almost no change to the waterway." Emrgy's turbines are designed to work in aqueducts and canals. The US federal government already has 15,000 miles of artificial waterways that could be generating local energy immediately.[29] For locations near waterways, distributed hydro promises to be a cost-effective, non-intermittent source of electricity. Coupled with the growing acceptance of placing solar panels over canals to reduce evaporation, waterways around the world can generate electricity in two ways at once, thus cutting the cost of wiring and electronics in half.

Local energy is building momentum

So much has changed since I decided to leave the technology industry in 2016 for the "greener" fields of clean energy. At that time, the microgrid I saw at the Pedro Fernandez school would have been prohibitively expensive, if it could have been built at all. Tesla's market-leading Powerwall 2 battery had only just been announced, solar panels were twice as expensive, and the expertise to design such a system barely existed.

What a difference a few short years makes!

- *Rooftop solar has doubled.* The number of US solar rooftops grew from 1.4 million in 2016 to well over 2 million by the end of 2019.[30] The US is on track to have a total of 3.4 million solar rooftops by the end of 2021.[31] PEW Research reports that almost half of non-solar homeowners are considering adding it.[32]

- *Community solar increased four-times.* Installations grew from 334 in 2016 to 1,466 at the end of 2020.[33]

- *Batteries have become mainstream.* Local energy systems with batteries have grown from a few thousand to more than 200,000 as of mid-2021.[34] (This has been largely constrained by availability.)

- *New projects everywhere.* Since 2016, larger local energy systems like solar microgrids have been appearing on islands, campuses, churches, military bases, schools, universities, office parks, municipal buildings, emergency services, and health care clinics across the US and the globe.

For all this good news, there is still a long way to go, and much of that involves our politicians at every level having the vision and the boldness to spur this innovation forward. It took the combination of a tragic hurricane and a failing electric utility to finally get Puerto Rico's governor to loosen regulations so that the Pedro Fernandez microgrid—and hundreds of others around the island—could be built. We need to do better. So far, only a small portion of the island's 3 million inhabitants have benefitted from these changes.

Puerto Rico offers invaluable lessons for the rest of the US and the world. Political leaders can remove the artificial barriers to local energy today rather than waiting for a natural disaster to force their hands. Communities around the world now have proven case studies on the resilience, environmental benefits, and cost savings thousands of Puerto Ricans are already enjoying.

As these systems increase and spread around the globe, I believe we will see Hurricane Maria as a tragedy that made possible lasting positive change, bringing about a local energy revolution that will help the world move toward 100% clean power.

〔〕 〔〕 〔〕

Local energy is already accelerating and changing every energy market it touches. It is the fastest, cleanest, easiest, most secure, and reliable way to build new electricity generation. It brings resilience, equity, and millions of good-paying jobs to the communities where it is built. And increasingly, local energy is the cheapest way to generate electricity. Because it lives outside the reach of the electric monopolies, genuinely competitive markets are emerging for the first time since the "War of the Currents" a century ago.

In the coming years, thousands of innovators and tens of billions of investments will be flooding in to propel the local energy revolution, producing a series of applications and improvements for local clean energy that will rapidly increase its commercial fortunes. Why am I so sure? The next three chapters will explore the underlying technology trends, the coming flood of innovations, and the disruptive forces they are unleashing.

CHAPTER 4

FROM FUELS TO TECHNOLOGIES

Thomas Edison had some extraordinary friends. Every summer toward the end of his life, he would assemble a few of the most celebrated men in America and set off into the wildness for extended trips. These expeditions into the still-wild areas of states like North Carolina, Virginia, Michigan, and New York were supported by several cars of staff and specialized vehicles bearing food, drink, and equipment including iceboxes, a kitchen stove, electric lights, and a folding table that seated twenty people. Maybe it was more "glamping" than camping, but these were the types of comforts that were expected by the giants of American industry who regularly made this trip, a group that included Edison's close friends, Henry Ford of Ford Motor Company, and Harvey Firestone of Firestone Tires.[1] They referred to themselves as the "Vagabonds."

What did these business visionaries talk about as they sat under the stars? One snippet survives in a memoir by James Newton, a businessman and friend of Edison, who recalled one of these conversations in his 1989 book, *Uncommon Friends*. Newton describes an exchange, perhaps inspired by their natural surroundings, of each man's thoughts on the energy demands being made by America's rapid, electricity-driven industrialization. How could a system almost entirely based on burning unrenewable resources, be sustained?[2]

FIGURE 4.1 *Henry Ford, Thomas Edison, President Harding, and Harvey Firestone discussing the day's news while camping. Credit: Smith Archive/Alamy Stock Photo*

Edison began, "We are like tenant farmers, chopping down the fence around our house for fuel, when we should be using nature's inexhaustible sources of energy—sun, wind, and tide."

Firestone responded that oil and coal and wood couldn't last forever . . . He wondered how much hard research was going into harnessing the wind, for example. "Windmills hadn't changed much in a thousand years."

Ford said there were enormously powerful tides . . .
Scientists had only been playing with the question so far.

Edison said, "I'd put my money on the sun and solar energy. What a source of power!"

"What a source of power!"

Even back then, Edison saw clearly what has taken the rest of the world a century to understand. There is no other source of energy like solar. It is not just that we will never run out of solar or that it leaves behind no wastes or pollution. What makes solar energy truly unique is the sheer amount of it.

Consider the amount of solar energy that bathed the planet in just one of those months that Edison and his friends spent in the American wilderness, July 1921. Over those 31 days in 1921, 60,000 exajoules of solar energy reached the surface of our planet—an incomprehensible amount of energy.[3] If the sun sent our planet an energy bill for that July, it would be for 16,790,000,000,000,000 kilowatt hours.[4] That is the amount of energy generated by 24 million nuclear power plants.[5] In just the first eight hours of July 1, the continents of the earth received more solar energy than all the world's 7 billion people used for all types of energy for the entire year of 2018.[6] In that single month of July, our planet received more energy from the sun than exists in all the combined reserves of oil, coal, natural gas, and uranium put together.

Was there anything unique about that July? No, it was a month like every other month for billions of years of Earth's history. Solar energy is not only boundless, it is also a feedstock for virtually every other type of energy we use today.

Hundreds of millions of years ago, primeval plants used photosynthesis to convert sunlight into carbon-based energy. Eons of geologic forces compressed those dead plants into what is today all the world's coal reserves. Oil and natural gas were also born from ancient sunlight, but in their case, marine microorganisms harvested the sun's energy instead of plants.

Every bit of wind that turns modern turbines is created from the solar energy that heats the air surrounding our planet. That wind, in turn, powers ocean waves. Hydropower dams tap flowing rivers, all of which are fed with rainwater evaporated by sunshine. Even nuclear fuels like uranium were created from a sun, although it was not ours. Born from supernovae billions of years ago, these dense, radioactive elements floated through space, eventually condensing with other elements to become earth (Figure 4.2 visualizes that much more solar energy is available compared with fuel-based and other renewable energy sources).

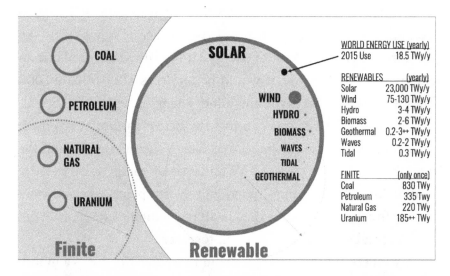

FIGURE 4.2 *The sun provides more energy to the earth each year than fossil fuels, nuclear, or any other type of renewable energy. Note: Geothermal and uranium have large potential, but it may not be technically and economically feasible to access it. Courtesy of Richard and Marc Perez.*[7]

As Elon Musk once put it, "We have a giant fusion reactor in the sky that works with no maintenance." And "That free fusion reactor conveniently converts ~4 million tons of mass into energy every second. We just need to catch an extremely tiny amount of it to power all of civilization."

Solar is not only abundant, but also the most equitable source of energy. Countries like Saudi Arabia, Russia, and the US may have won the geologic lottery for fossil fuels, but solar is the true equalizer. It shines down freely upon every part of the world, from the wealthiest nations to the poorest villages. As President Jimmy Carter once said, "No one can ever embargo the sun." Every region and every country receive more than enough solar energy to power all their homes, their cars, and their industries—everything.

So how is it that the most abundant, and least expensive form of energy made up just 2% of US electricity in 2020?[8] That is the trillion-dollar question and one that I will illuminate shortly and throughout the book. But before I do, let us take a stroll through the history of solar and how we ended up in this paradoxical situation.

A SHORT HISTORY OF SOLAR

The ancient Greeks, Romans, and Chinese were among the first cultures to tap solar energy. By building their houses facing south and placing windows in the right locations, they used sunshine to heat their homes during cold winter days. It would be another few thousand years before the first devices were built to directly harness sunshine. As John Perlin explains in his book *Let It Shine: The 6,000-Year Story of Solar Energy*, a heightened interest in personal hygiene in the late 1800s drove the popularity of taking hot baths. For years, wood fires were the only way to heat the water. But late in the century, the first solar water heaters began appearing on the market, replacing the laborious and expensive process of burning wood.

Modern solar water heaters consist of a sun-facing box filled with black pipes that absorb the sunlight and heat the water flowing through them. While not widely embraced in the US today, solar water heaters remain popular in countries like Austria, Cyprus, Israel, and Greece.

In the first half of the 20th century, the rapid ascension of electric energy and the Big Grid relegated the solar water and home heating industry to a small niche. It was not until the 1980s that solar innovation had developed enough to play a role among the giant coal, hydro, and nuclear power plants that dominated the grid.

The first type of large-scale solar plant was *concentrated solar power* (CSP), which uses mirrors to concentrate sunshine in order to super-heat a fluid, which then spins a turbine to generate electricity. The most common CSP design is called a *solar thermal tower*—acres of moveable mirrors track the sun across the sky, focusing its energy on a receiver sitting atop a giant tower. CSP plants have one big advantage over most other types of solar plant: the heated fluid can be stored for as long as a day and released as needed, offering solar power even when the sun is not shining.

There was a lot of initial excitement over CSP when the technology was first commercialized in the 1980s, and dozens of plants were built across the world.[9] Unfortunately, even though the large, centralized CSP plants were a comfortable fit with the financial and technical models of the century-old grid, their costs never declined enough to become competitive. In 2020, the cost of electricity from a new CSP plant was 10.6 cents per kilowatt hour, higher than even nuclear's 9.9 cents.[10]

As CSP struggled with operational and cost challenges, another very different way to capture solar energy had been quietly improving in efficiency and cost. Around 2011, as CSP lost its early cost advantage, this new technology not only overtook CSP, but it soon became the fastest growing method of generating electricity in the world.[11]

The alchemy of transforming sunlight directly into electricity

Science fiction author Arthur C. Clarke once said, "Any sufficiently advanced technology is indistinguishable from magic." These words seem an apt description of what is now the most widespread method of harnessing solar energy—*photovoltaic solar*, or PV. Rather than relying on heat as an intermediary form of energy, PV directly converts the energy in light into electricity. It is like an alchemy that transforms photons into electrons. To understand how this near-magical ability became mainstream, we need to go back in history a few hundred years.

The scientific basis for photovoltaics was discovered in 1839, when the French scientist Alexandre-Edmond Becquerel, at the young age of nineteen, demonstrated that light striking certain materials created an electrical current—enshrining another term for PV still referred to today as the "Becquerel effect." Building on this work, two British scientists, William Grylls Adams and Richard Evans Day, discovered in 1876 that the element selenium was particularly adept at this conversion of photons into electrons.

This discovery was first put to commercial use about a decade later by an American inventor named Charles Fritts, who produced the world's first roof-mounted photovoltaic (PV) panels in New York City. This was in 1884, two years after Edison flipped the switch at Pearl Street Station. Now humanity had a system that could generate an electric current on the same building that used it, with nothing more than sunlight! No need for power plants belching black smoke, coal mines, costly fuel, or dangerous overground distribution lines.

Unfortunately, Fritts's pioneering panels required expensive selenium and gold and converted only 1%–2% of solar energy (modern panels convert 18%–22%). While many at the time yearned for an electricity source that required no fuel or fire, Fritt's panels were ahead of their time and today are virtually forgotten. With powerful coal plants proliferating across the country in the 1880s, solar never had a chance as a serious competitor.

FIGURE 4.3 *Charles Fritts installed the first solar panels on a New York City rooftop in 1884. Courtesy of John Perlin.*

The next big leap in solar took place not in a lab or a roof, but in one of history's greatest minds. Albert Einstein may be most remembered for his theory of relativity and his iconic formula, $E=mc^2$, but it was his discovery of the physics underlying the direct conversion of light into electricity that won him the Nobel Prize in 1921.[12] His work laid the theoretical foundations for the science that powers the solar industry today.

It was another 30 years before that science began its journey toward commercial viability. At a press conference in April 1954, Bell Labs announced it had created a silicon-based solar cell that handily beat the 2% efficiency barrier. A *Newsweek* report called solar energy ". . . an eventual competitor to atomic power."[13] But again, PV was still far from being cost competitive with established power sources like coal, natural gas, and nuclear plants.

In 1956, one of the scientists responsible for the breakthrough, Daryl Chapin, highlighted the cost gap by calculating how much money it would take for a typical homeowner to buy an array that could power their house. Back in 1956, the cost was $1,430,000— almost $14 million in current dollars.[14] The road in front of solar would prove to be long, but the journey had begun.

Solar finds its first commercial market in space

In the 1950s, there was one buyer with sufficiently deep pockets to pay for solar: The US government. Bell Labs' solar cells turned out to be pivotal to the burgeoning US space industry. Money would become even less of a problem after October 4, 1957, when the Soviet Union stunned the American public by launching the first artificial satellite into orbit, the Sputnik 1.[15] American complacency about its technology superiority was shattered as the Soviet's satellite radio signals were picked up by shortwave radio hobbyists around the globe. Sputnik mesmerized the world for 21 days—until its radio signals went silent after its onboard batteries died.

On January 31, 1958, America launched its answer to Sputnik, the Explorer 1. This was a more sophisticated satellite equipped with scientific instruments that, among other things, led to the discovery of the Van Allen radiation belt surrounding the Earth. But even with its superior batteries, it too fell silent after four months.[16] It was America's next successful satellite launch, on March 17, 1958, that changed the nature of the space race. This satellite, named Vanguard 1, had solar panels. The satellite maintained its radio signals for six long years, offering the world its first glimpse of sustainable solar power. It still circles Earth today. No other human-made object has been in space longer.

Oil companies bring solar down to earth

Ironically, it was an oil company's research that helped make solar affordable for Earth-based applications. In 1973, Exxon created a subsidiary called Solar Power Corporation to solve a very real problem—finding an alternative to the expensive batteries that were then used in areas too remote to be reached by the grid. In his book *Let it Shine*, John Perlin explains that Exxon developed a new manufacturing approach that reduced the cost of silicon solar far below the cost of cells used in space. The first application of these more affordable cells was to replace the expensive one-time batteries that powered the foghorns and lights on Exxon's offshore drilling platforms. They quickly found new markets for these cells, for use in ocean buoys and remote communications repeaters. Exxon eventually shuttered the business around 1984, concluding it would be decades before it was profitable.[17]

Taking solar global

Perhaps no one was more important to solar's proliferation in these early years than Martin Green, an Australian scientist who, like some globe-trotting Johnny Appleseed of PV panels, introduced the technology to countries around the world. Green first became interested in solar in the '70s while he was in the United States, then the leader in both the scientific and manufacturing parts of the solar industry. As Greg Nemet describes in his book, *How Solar Energy Became Cheap*, Green brought a new network and eventually used American PV manufacturing equipment back home to Australia and began small-scale production in the early '80s. Eventually he was teaching at the University of New South Wales's School of Photovoltaic and Renewable Energy Engineering (SPREE). One of his students, Shi Zhengrong, returned to China and established Suntech, which for a time led the world in panel production, making Shi briefly the richest man in China. Today, Suntech is just one of many Chinese solar manufacturers, which together produce 70% of the world's solar panels.

The solar industry has come a long way from the early days of solar-powered satellites to millions of modern-day panels shipping from Chinese factories. One of the most remarkable parts of industry's evolution is the unprecedent declines in cost—over the last 50 years, the price of solar has dropped 400 times. Understanding how this has happened is one of the most important stories in this book.

SOLAR IS A TECHNOLOGY, NOT A FUEL

A few years before I was born, my father had a brief stint as a computer programmer, working on the last operating UNIVAC computer. This groundbreaking machine was one of the first computers to be used primarily for business. It weighed 8 tons, had 5,000 vacuum tubes (the precursor to transistors), cost upwards of $10 million in current dollars, and could calculate 1,900 operations per second.[18]

FIGURE 4.4 *Inside a UNIVAC 1 computer. Robert Edward Nussey (1938–1993) is second from the right. Source: Bill Nussey.*

By comparison, I am typing these words on an Apple MacBook Air computer that weighs 3 pounds, has 16 billion transistors, costs $1,000, and can perform 36 billion operations per second.[19] Computers have come a long way.

Gordon Moore, one of the founders of microchip giant Intel, famously forecast the early progress of computers in 1965, in what is surely the most famous article ever published by the trade publication, *Electronics* magazine. His prediction, now known as Moore's Law, simply states that the number of transistors on a microchip will double every two years.[20]

As Moore explained on the 50th anniversary of the original publication:

> Oh, I'm amazed. The original prediction was to look at 10 years, which I thought was a stretch. This was going from about 60 elements on an integrated circuit to 60,000—a thousandfold extrapolation over 10 years. I thought that was pretty wild. The fact that something similar is going on for 50 years is truly amazing.

Intel's CEO at the time, Brian Krzanich, put this in perspective saying that if Moore's Law applied to a car, "you would be able to go with that car 300,000 miles per hour. You would get two million miles per gallon of gas, and all that for the mere cost of 4 cents!"

My laptop is testament to both the vision of Gordon Moore and the near magical ability of scientists to endlessly continue squeezing more value onto a wafer of silicon. Having spent most of my career in the computer industry, it has been inspiring and humbling to see how rapidly the industry has evolved. New products and new technologies roll out daily. One year, a new product reshapes business. A few years later, it is obsolete.

The experience curve

No other industry can match the relentless pace of computers and digital communications, but the fundamental principles of technology progress are fairly universal. Bruce D. Henderson, the founder of the iconic strategy consulting firm, Boston Consulting Group (BCG), called one of the principles that underlie technology progress the *experience curve*. It describes how, as manufacturing volumes grow, organizations gain experience, which translates into reduced costs. In other words, the more you make, the cheaper it gets.

The experience curve can also be applied to the electricity industry, but its effect is more limited. There are two reasons for this. First, reducing electricity costs is best achieved through *economies of scale*, not high-volume mass production. In other words, the best way to lower electricity costs has always been to build a smaller number of larger and larger power plants.

The economics of nuclear power plants epitomize this. At 1,000 megawatts per plant, these are not only the largest type of power plant, but also some of the largest and most complex machines ever built by humankind. The problem is that larger plants—which can take a decade or longer to construct—mean fewer are built and there is far less opportunity to put the experience curve to work.

Consider how many power plants the world has built. At the end of 2019, there were 3,900 natural gas plants, 2,400 coal plants, and just 440 operating nuclear power plants.[21] It should not be surprising that natural gas plants have seen the largest improvements in efficiencies and costs and that nuclear plants have seen almost none.

The second reason the experience curve has little effect on the traditional electricity business is fuels. The biggest cost for fuel-based electricity generation is the fuel itself. Economies of scale and the experience curve stopped having a major impact on the price of fuels decades ago. Today the price of fuels is more affected by supply and demand and government policies than any technologies that supply them (the commercialization of fracking natural gas being the major exception).

In the late 20th century, wind turbines were the one type of "big machine" that was able to break these economic constraints and benefit from both scale and manufacturing experience. The average size of a new wind turbine is growing larger year over year, driving improved economies of scale. But because each wind turbine is relatively small compared with fuel-based power plants, there are a lot more of them. Many parts of wind turbines are manufactured in factories and shipped onsite for assembly. With approximately 500,000 wind turbines in operation today—one hundred times more than other power plants—these machines are also benefiting from the experience curve.[22]

The new economic principle reshaping energy: economies of volume

The electricity business has driven down costs over the last hundred years primarily thanks to economies of scale—making a small number of really large things (power plants). The computer industry, on the other hand, has driven down costs by making a huge number of smaller and smaller things—in other words, computer costs have benefitted from what I call the *economies of volume* that come through mass production.

Every aspect of manufacturing a computer or a mobile phone can be improved as more of them are manufactured. Factories get larger. Automation reduces expensive labor. Precision increases. Quality improves. Engineering costs are spread over a higher number of units. Global supply chains lead to massive purchasing efficiencies. All of this creates a virtuous cycle that delivers a stream of better, cheaper products to customers.

How do economies of volume apply to solar? The basic unit of solar power is very small—a solar cell. Most modern cells are thin slices of pure crystalline silicon with various materials and processes applied to make them more efficient and durable. Most are about 6 inches square and as thin as two human hairs. The industry has made a lot of them, *more than 120 billion individual cells!*[23] Whether a cell is used by itself to power a portable lantern in Africa or used along with millions of others to power a city, individual cells are all the same.

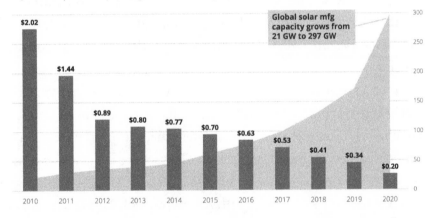

Solar PV costs decline as manufacturing expands
(US$ costs per watt in panels / global manufacturing capacity in gigawatts)

FIGURE 4.5 *The learning curve effect and economies of volume cause the cost of silicon solar PV to fall as manufacturing capacity increases. Sources: Clean Energy Associates, International Energy Agency, Our World in Data (freeing energy.com/g201).*

Compared with mere thousands of power plants, and hundreds of thousands of wind turbines, it is easy to see that manufacturing billions of solar cells will achieve a unique degree of learning and economies of volume. This is why the costs of solar have plummeted more than 400 times since the early days of space satellites, when solar cost $76.00 per watt. Over the last ten years, factory capacity has expanded more than 10 times to nearly 300 GW, driving the cost of solar from $2.02 down to $0.20 per watt (see Figure 4.5). As the clean-energy expert Michael Liebreich has written, "A solar panel is essentially a flat-screen TV, but a lot simpler. Why wouldn't its costs behave in a similar way?"

If comparing solar cells and panels to a nuclear plant strikes you as odd, you can start to see why the electricity industry struggles to understand the new economic principles reshaping their industry.

The Department of Energy is "calling the shot" on solar and batteries

During the fifth inning of Game 3 of the 1932 World Series, baseball legend Babe Ruth stepped up to the plate, pointed to the outfield, and then hit one of the most famous home runs in history. Calling the shot, especially when odds are long and the crowds are rowdy, transforms predictions into inspirations.

Another, somewhat less legendary but equally inspiring, "call the shot" occurred in 2011 when the US Department of Energy (DOE) launched the SunShot Initiative. At the time, they said utility-scale solar cost 27 cents per kilowatt hour. They set the audacious goal of driving costs down to 6 cents by 2020.[24] Their "call the shot" was Babe Ruth-like and was met with no shortage of skepticism across the industry. But the DOE pushed forward, investing hundreds of millions in research and pilots. They worked with businesses to add billions of dollars of private capital toward the goal. Not only did solar hit that incredible 6 cent target, but it achieved it in 2017, *three years ahead of schedule.*

The DOE is once again setting a bold target for driving down the cost of solar. In 2021 it set a new goal of lowering large-scale solar from 4.6 cents per kilowatt hour in 2020 down to 2 cents by 2030! This time the DOE is also setting goals for local energy scale solar projects, too—with residential costs dropping from

13 cents per kilowatt hour in 2020 down to 5 cents by 2030, and commercial costs dropping from 9 cents to 4 cents in the same time frame.

SunShot targets from the Department of Energy
(levelized cost of solar PV electricity in cents per kilowatt hour)

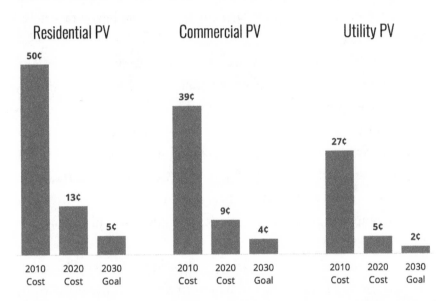

FIGURE 4.6 *The progress and most recent targets from the Department of Energy's SunShot Initiative, which strives to help dramatically lower the cost of installed solar. Source: Goals of the Solar Energy Technologies Office, US Department of Energy.*

Not content to focus merely on solar, the DOE launched the Energy Storage Grand Challenge in 2020.[25] This initiative sets a new set of audacious targets. For long-duration storage, the DOE is targeting a 90% reduction in cost by 2030. For mainstream lithium-ion batteries, the DOE's 2030 goal is $80 per kilowatt hour, half of its estimated costs in 2020. Honestly, I think the innovators in these markets will beat all these targets, just as they did the 2011 goals.

Why silicon is winning the PV race

Over the years, many different materials have contended for the cheapest type of photovoltaic solar cell. While the race is still under way, silicon had risen to 95% of the market in 2020 and is likely to remain the primary material used for solar power over the long term. How has silicon managed to pull ahead amid a host of competing materials? Because silicon solar cells had a large head start down the experience curve.

Think back to the story about the breathtaking progress in the computer industry. It turns out that solar cells are remarkably similar to microchips. When the PV solar industry began its explosive growth, it was able to leap forward decades on the experience curve of working with silicon, thanks to borrowing the machines, the processes, and the supply chains that Gordon Moore and other chip pioneers had laid down years earlier. As one solar scientist told me, "We know more about the physics and chemistry of silicon than any other material on earth."

Our deep understanding of the properties of silicon is one of the reasons the solar industry has been able to come up with its own version of Moore's Law. It is called Swanson's Law, after Dick Swanson, the founder of pioneering solar company, SunPower. The law predicts that the cost of solar will drop 20% with every doubling of manufacturing capacity. Like Moore's Law, Swanson's prediction has proven surprisingly accurate since it was first posited in 2012.

Fuels are incremental. Technologies are disruptive.

All this discussion about laws of volume and scale, Swanson and Moore, is merely backdrop for one of the most important facts I want to impart about solar; a simple truth that makes solar different from any other energy source before it: *solar is a technology, not a fuel.* Once you understand this distinction, you will understand the first of solar's four unique "superpowers" and why solar is disrupting the energy industry. More on superpowers shortly.

Why is being a technology so advantageous? For a century, most of our electricity has been generated from fuels. Every kilowatt hour sent onto the grid required another batch of fuel to be consumed. Fuels must be mined or drilled. They must be delivered to power plants, often through complex global supply chains. Once a batch of fuel is used up, it is gone forever.[26] In its place, we are left with some form of residual waste, like greenhouse gases, coal ash, or radioactive waste.

Every step of the fuel life cycle has profound environmental impacts and increases the ultimate cost of the kilowatt hour. Even if we avoid paying for these costs in our current electric bills today, the full costs will come due to our children and grandchildren in future decades. Renewables like solar, wind, and geothermal change the rules and the business models because they require no fuels. They suffer few of the costs and consequences of fuel-based generation.

Moving beyond a fuel-powered grid toward renewables is an important step. But only solar takes the final step. In an industry full of giant machines, only solar is pure technology. Cells and panels are made by the billions in vast, highly automated factories. Each cell manufactured improves the experience curve. Economies of volume make each new generation cheaper than the last. Photovoltaic solar has more in common with your phone or laptop than it does with the giant thermal power plants dotting the world. Once manufactured, solar products are incredibly simple. The vast majority of installations have no moving parts.[27] This is why photovoltaic solar has dropped faster than any other type of electricity generation (see Figure 4.7). And unlike other types of power plants, the cost of solar is independent of unpredictable fuel costs, making it steady and predictable across the lifetime of the solar installation.

Changes in the cost of generating electricity over a decade
(levelized cost of energy (LCOE) in cents per kilowatt hour from 2010 to 2020)

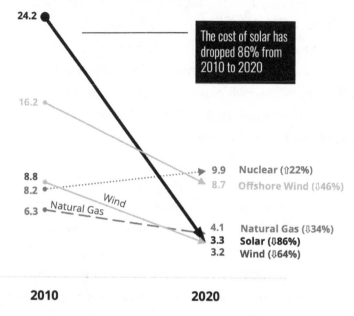

FIGURE 4.7 *Between 2010 and 2020, the cost of generating electricity from solar has dropped further and faster than the costs declines of natural gas, solar, onshore wind or offshore wind. Sources: EIA, NREL, LBNL, Wood Mackenzie, BNEF, and Lazard (freeingenergy.com/g210).*

Looking forward, solar will be cheaper than natural gas and even wind in the coming years (see Figure 4.8). Virtually every independent research group and analyst is aligned on the inevitability of solar becoming the cheapest way to generate electricity ever created.

What is surprising is that very few policy makers, utility executives, and grid designers seem to be acting on this inevitable outcome. Why is this?

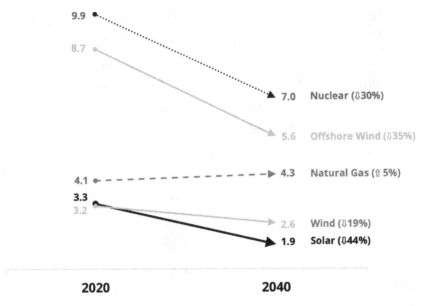

Solar will be the cheapest way to generate electricity
(levelized cost of energy (LCOE) in cents per kilowatt hour from 2020 to 2040)

9.9

8.7

7.0 Nuclear (⇩30%)

5.6 Offshore Wind (⇩35%)

4.1 ← → 4.3 Natural Gas (⇧ 5%)

3.3
3.2

2.6 Wind (⇩19%)

1.9 **Solar (⇩44%)**

2020   2040

FIGURE 4.8 *Every research group predicts solar will be cheaper than wind, natural gas, and nuclear by 2040. Sources: EIA, NREL, LBNL, Wood Mackenzie, BNEF, and Lazard (freeingenergy.com/g211).*

SOLAR—THE RODNEY DANGERFIELD OF ENERGY

Whenever I puzzle over the energy industry's struggle to embrace solar, I am reminded of the famous gag line of comedian Rodney Dangerfield: "I get no respect, no respect at all." (Feel free to look this up if you are too young to know the reference.) It is difficult to find a bigger underdog in the energy industry than solar. While this is slowly changing, Chapters 7 and 8 explore just how many uphill battles remain.

Why is solar marginalized, disrespected, and generally slow rolled? I think there are two reasons. The first is that solar is evolving so quickly that few people can make sense of it. Hockey legend

Wayne Gretzky famously said the key to winning is to skate where the puck is going. If the puck is the price of solar, then the electricity industry is like one of those Zamboni ice-cleaning machines—it moves too slowly to have any chance of keeping up with the puck, let alone skating to where it is going. This comparison is lighthearted, but the consequences are very real. The foremost agencies in the world tasked with providing data for grid planners are consistently and woefully undershooting the mark on solar.

The US Energy Information Administration (EIA) is a branch of the Department of Energy that plays a critical but behind-the-scenes role in the nearly $1 trillion US energy industry. The EIA collects and publishes an impressive range of data—from the daily price of natural gas to the average height of wind turbines. All the information is freely available, and it is the most common source of the data used in this book. But there is one area where the EIA really falls short. It is terrible at forecasting the growth rate of solar.

In 2005, the EIA forecast it would take 20 years before the US would have 0.40 gigawatts of total solar capacity.[28] That number was beaten handily. By 2009, just four years later, the US had 0.53 gigawatts of solar—16 years ahead of the EIA forecast.[29] The EIA continued wildly low-balling solar year after year, and it has only recently embraced a more realistic growth model for solar. This pessimism about solar is not necessarily some US conspiracy. The EIA's international sibling, the International Energy Agency (IEA), has also famously under-projected solar for years.[30]

How can these organizations be so wrong so consistently? The EIA published an open letter in 2016 defending its projections, pointing to unpredictable policies as the biggest source of its errors.[31] No doubt that is a component, but given the substantial and ongoing disparities, I believe the largest culprit was its inability to understand the cost curves of a technology like solar. We will never know the

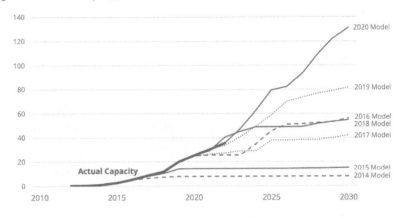

Installed US solar capacity: government projections vs reality
(Gigawatts of total solar capacity)

FIGURE 4.9 *Government agencies like the US Energy Information Administration (EIA) and Europe's International Energy Agency (IEA), which have consistently underestimated the growth of solar, only recently started reflecting solar's rapid growth. Source: EIA's Annual Energy Outlook 2014-2020 (freeingenergy.com/g126)*

full impact of these woefully conservative predictions, but it is a good bet that countless decisions would have been made differently if the EIA and IEA had embraced the reality of technology-driven price declines that the computer and communications industry mastered decades ago.

There is a second reason I believe solar "gets no respect." It democratizes energy. Or, as I like to say, *solar is freeing energy.* It unlocks energy from giant, centralized systems controlled by governments and corporations. It shifts energy profits from Wall Street to Main Street. It offers new depth and meaning to the phrase "power to the people." It upends nearly every economic, techni-cal, and policy assumption underpinning the multitrillion dollar energy market. In short, the idea that anyone can generate their own energy is a serious headache for some of the world's most powerful institutions.

How can any single kind of energy production have such a large impact on a world dominated by powerful incumbents like fossil fuels providers and utilities? It is because solar has superpowers that no other type of electricity generation has ever had before.

THE SUPERPOWERS OF SOLAR

In the 1952 TV show, *The Adventures of Superman,* America's favorite superhero had four superpowers. He was faster than a speeding bullet and more powerful than a locomotive. His power of flight meant he could leap over tall buildings in a single bound. And his superpowered eyes gave him both X-ray and heat vision.

A bit like our comic book hero, solar has four superpowers that make it more powerful than any energy generation system before it. We have already covered the superpower #1: solar is a technology (not a fuel).

The second and third superpowers—*dual-use* and simplicity—are also unique to solar and equally game changing. The last superpower is the most unique and the reason solar is the superhero of this book. Solar is local.

Solar superpower #2: dual-use

Solar can piggyback on existing surfaces and make otherwise mundane roofs, walls, windows, roads, and lakes into miniature power plants. This is a "1+1=3" synergy.

The most widespread example of dual-use is good old rooftop solar, but this is just the beginning. Solar projects are revitalizing land that was thought to be unusable. Landfills, coal mines, Superfund sites, and other *brownfield* locations are hosting solar projects and bringing jobs and tax revenues into locations that were otherwise off-limits. Even the site of the world's worst nuclear accident, the Chernobyl nuclear plant, is now home to a 4-acre solar farm.[32]

Solar can be added to existing structures in an endless variety of ways. Georgia and France have both piloted replacing roadway asphalt with solar panels. A Dutch contractor is testing highway noise barriers made of solar panels. A 20-mile-long bicycle lane in South Korea generates solar power as it shields bikers from the direct sun. Two German companies are mounting panels on the sides of giant grain silos. India is covering canals with solar panels to make better use of the space and to reduce the evaporation of water.[33] These applications are bleeding edge, so the jury is still out on their cost-effectiveness. But there are five additional dual-use solar opportunities that are so large and economically viable that, taken together, they can provide almost all the electricity consumed by the US.

Residential rooftops. A 2016 study by NREL found that small buildings, which are largely residential, could host enough rooftop panels to generate 926,000 gigawatt hours of electricity per year, about 23% of all US electricity needs.[34]

Commercial building rooftop. The same study found that larger buildings, which are generally commercial and industrial, have the potential to generate 506,000 gigawatt hours per year, about 13% of what the US consumes each year.

Floating solar. Also called *floatovoltaics*, solar panels are installed on racks that float on the surface of water bodies. One study found that adding floating solar to just a quarter of human-made water bodies would produce 786,000 gigawatt hours of electricity a year, about 20% of all US electricity consumption.[35] Not only does the proximity to water help cool the panels and make them more efficient, but the panels reduce evaporation and limit algae blooms.

Canopy solar on parking lots. With an estimated 3,600 square miles[36] of parking lots in the US, overhead solar panels not only protect cars and people from the elements, they also have the potential to provide 20% of US power needs.

Cropland and grazing. Chapter 8 describes an exciting new idea called *agrivoltaics,* that is, intermingling solar and agriculture on the same land. Solar panels actually improve the growth of livestock and certain crops by shielding them from direct sunlight and reducing evaporation.

The point here is not that all these dual-use applications need to achieve their full potential, only that there are nearly endless locations for solar that do not require replacing pristine fields, farms, or forests with solar panels.

Solar superpower #3: simplicity

Solar is, by far, the simplest way to generate electricity. You take the panel out of the box, mount it facing the sun, wire it up, and you are generating electricity. Maintaining solar is even simpler. You periodically clean the panels and replace the occasional failed part. Of course, manufacturing solar panels is a lot more complex, but even this is simple compared with making a gas turbine or a nuclear reactor.

Centralized power generation is complex

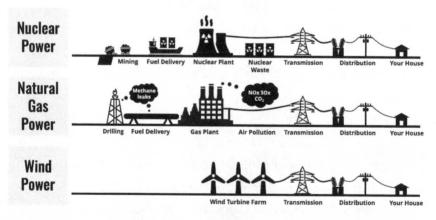

FIGURE 4.10 *Most large, centralized power plants rely on complex fuel supply chains and require some mechanism for disposing of wastes. Even clean energy like wind relies on complex transmission and distribution to deliver electricity to homes.*

Unlike solar, traditional centralized power plants like nuclear and natural gas rely on a complex dance of fuel supply chains and electricity distribution (see Figure 4.10). Fuels like uranium and natural gas need to be extracted and delivered to a plant. The plant uses the fuel to heat water or air, which then turns a generator to produce electricity. Consuming the fuel leaves behind waste. Natural gas combustion wastes are all gases, including greenhouse gases, most of which can be released into the air with no additional financial costs. However, if politicians decide a carbon tax is the best approach for reducing greenhouse gases, then the cost of natural gas-powered electricity will go up—a lot. Nuclear waste is far more regulated, so the industry already has a complex and expensive multi-tiered process for handling waste.

The electricity generated at these centralized plants is fed into the high voltage transmission grid where it travels as many as hundreds of miles. Substations are giant transformers that lower the voltage and send the electricity out on the distribution grid, which ultimately delivers it to a house or a building. Large, centralized renewables like wind skip the entire fuel supply and waste-handling cycling but still rely on millions of miles of transmission and distribution power lines.

Like wind, large-scale solar is much simpler than fuel-based power plants, but they both remain reliant on transmission and distribution. It is when the solar panels are moved from giant farms to local rooftops that solar's simplicity begins to shine (see Figure 4.11). These local energy systems still require access to the grid at night and on cloudy days, but the economic benefits of solar now accrue directly to the customers that own them. Solar's true promise for simplicity emerges when it is paired with batteries. These local energy systems have the potential to be entirely stand-alone, with no fuel supply chains, no wastes to clean up, and no reliance on complex grids.

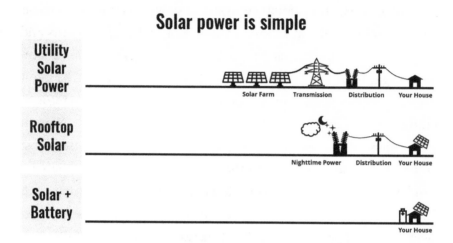

FIGURE 4.11 *Local solar reduces the reliance on some of the most complex aspects of centralized power plants.*

Why is simplicity important? Simple systems are easy to deploy, have lower risk, and most importantly, they can be installed by anyone with modest training, creating one of the largest job opportunities in the US. You just install a system, and it sits there for decades, silently generating electricity at virtually no cost. Solar installations have few single points of failure because they are made with large numbers of identical parts. Like a stadium, if a few of the seats break, you just move people around and replace the seats later. You never have to shut the whole thing down.

Solar superpower #4: local

It is the superpower of simplicity that democratizes solar—anyone, anywhere can generate their own electricity. The vast majority of the world's population can afford to own a system, even if many are very small. This simplicity is also an essential element of solar's fourth and final superpower: that with local production, solar can be operated anywhere, by anyone and everyone.

This subject requires its own chapter, so for now, let us first look at several other critical technologies that complement solar—batteries, power electronics, and efficiency—and how they are enabling solar to flex all its superpowers.

BATTERY SUPERPOWERS

Find any list of history's greatest inventions and you will likely see the printing press, the electric light, the telephone, transistors, and antibiotics. But there is one invention that is often missing from these lists that I believe has also transformed society, albeit a bit less directly. In fact, you probably own many of them, including the one powering the phone in your pocket right now—the lightweight, rechargeable battery.

Alessandro Volta takes on frog power

In the late 1700s, Italian scientist Luigi Galvani believed he had discovered a new source of electricity, recounts Steve LeVine in his book *The Powerhouse: Inside the Invention of a Battery to Save the World*. Galvani had been dissecting frogs and noticed their legs would twitch when he probed them with different metals. He also noted the legs would jump when the frogs were hung on a metal fence during a lightning storm.[37] Based on this, he concluded that frog muscles were a new source of electricity. He called it "animal electricity."

Fellow scientist Alessandro Volta disagreed. He set out to prove Galvani wrong. Volta theorized that the leg muscles were reacting to electricity, not generating it. He believed it was the different metals in the probes in the presence of a liquid environment that was generating electricity. After years of experimenting with various metals, he finally found a successful combination. "Alessandro Volta invented the first battery and thus launched the electric age in 1799," LeVine proclaims in his book.

Over the next 120 years, batteries have powered everything from emergency lighting and portable radios to their most common application, powering the electric starters on gasoline-powered engines. But it was not until the invention of an affordable, lightweight, long-lasting, rechargeable battery—lithium-ion—that batteries truly changed the world.

The Nobel Prize-winning battery

On October 9, 2019, John Goodenough, Stanley Whittingham, and Akira Yoshino were awarded the Nobel Prize in Chemistry. Writing of their achievement, Sara Snogerup Linse of the Nobel Committee for Chemistry explained, "We have gained access to a technical revolution. The laureates developed lightweight batteries with high enough potential to be useful in many applications—truly portable electronics: mobile phones, pacemakers, but also long-distance electric cars. The ability to store energy from renewable resources—the sun, the wind—opens up for sustainable energy consumption."[38]

Recognition for their work was a long time coming. It began in the early 1970s, with Exxon once again playing a pioneering role. The company hired Dr. Whittingham to commercialize his research work in tapping the high energy potential of lithium to create a new generation of electric battery. After years of secrecy, he filed a patent in 1976 for a lithium-ion battery.[39] Alas, "exploding laboratories and disintegrating batteries . . . plagued Whittingham's work," explains LeVine. Exxon eventually shuttered the project.

A few years later, the most colorful of these three pioneers, John Goodenough, took lithium-ion batteries one big step forward. As the head of Oxford University's Inorganic Chemistry Lab, he performed breakthrough work in the lithium-cobalt-oxide cathode, making Wittingham's original idea far safer and more effective, leading to the world's first practical lithium-ion batteries. He was 97 years old when he became the oldest ever winner of the Nobel Prize, a long overdue acknowledgement of the world-changing impact of his work.

Akira Yoshino took the final step toward the mass production of lithium-ion batteries in 1985 by adding an anode made of graphite, making them safer and more commercially viable. A few years later, in 1991, the lithium battery humbly stepped into our lives by powering a Sony handheld video recorder. It was a blockbuster. Within a few short years, nearly all portable cameras were powered by lithium-ion batteries. But that was just the start.

Today, lithium-ion batteries can be found in mobile phones, laptop computers, smart watches, remote controls, e-book readers, portable video games, cordless power tools, and cordless vacuums. LeVine summed up their utility neatly: "The lithium-ion battery gave the transistor reach. Without it, we would not have smartphones, tablets or laptops. . . . There would be no Apple. No Samsung. No Tesla."

These are just the direct beneficiaries. Without the products made possible by the lithium-ion battery, there would be no ride-sharing apps, GPS-navigation, streaming music, audio books, or mobile payments, to name just a few.

Freeing transportation from its hydrocarbon chokehold

Billions of portable electronics devices led to massive economies of volume for lithium-ion batteries. The resulting low prices teed up these batteries for their most disruptive role yet—electric vehicles. While Tesla Motors has become synonymous with EVs, the company was far from the first to make an electric car. In fact, a century ago, batteries and gasoline engines were in a neck-and-neck race for auto supremacy. In 1900, battery-powered cars accounted for one-third of the entire market.[40]

One of the biggest proponents of electric vehicles back then was none other than Thomas Edison. "Electricity is the thing. There

FIGURE 4.12 *Thomas Edison holding the nickel-iron battery he hoped would power the future of electric cars. Credit: Scientific American, January 14, 1911.*

are no whirring and grinding gears with their numerous levers to confuse . . . no dangerous and evil-smelling gasoline and no noise," said Edison, according to Paul Hawken in his book *Drawdown*.

In 1903, Edison launched the Edison Storage Battery Company to commercialize a nickel-iron battery, which was lighter and better suited to vehicles than lead-acid. According to Russ Banham in his book *The Ford Century*, Edison's protege, Henry Ford, was so convinced of Edison's vision that he purchased 100,000 Edison batteries to commercialize an electric vehicle for Ford Motor Company. Alas, nickel-iron batteries suffered from intractable shortcomings—like operating poorly in cold weather. The project was eventually cancelled, and the EV revolution was postponed as it went in search of a better battery.

Hopes for a battery-powered vehicle revived decades later. In 1966, Ford Motor Company announced the EV revolution had finally arrived. This time it would be powered by its new sodium-sulfur battery, which promised 15 times higher energy density than lead-acid.[41] But operating temperatures of 300 degrees and other challenges posed by this battery delayed the EV revolution yet again.

Enter General Motors in 1996, with its all-electric EV1. The company leased over 1,000 of these cars. Initially powered by lead-acid batteries and upgraded to nickel-metal-hydride (NiMH) batteries, they were very popular and might well have sparked the long-awaited EV revolution. But in a surprise move, GM cancelled the EV1 in 2002. The company recalled every vehicle, angering customers and spawning countless conspiracy theories that linger today.[42]

It took a few more years, but the EV revolution did finally arrive. In 2009, Tesla Motors began shipping its first model, the Roadster. Then in 2010, Nissan shipped the more affordable LEAF. Tesla followed up again, by shipping the luxury Model S in 2012. EV sales started to take off.

How did Tesla solve the intractable problem of inventing a battery that could take a car as far as a gas tank? To the incredulity of auto experts, Tesla sidestepped a new battery chemistry entirely. Instead, it took the radical approach of wiring together thousands of small, commonly available lithium-ion batteries— the same kind used in consumer electronics. In one move, Tesla tapped into an existing, competitive worldwide supply chain. Tesla's strategy of using an already-mature supply chain shaved years off the time it took competitors to reach economies of volume. It is a wonderful irony that a company named after one of Edison's greatest rivals ultimately delivered on his vision of a mass-market electric car.

Hundreds of billions of batteries

As nearly every automobile manufacturer in the world now steers its gas-powered companies toward an electric future, the demand for lithium-ion batteries is soaring. Governments and companies across the world are investing billions in a breakneck race to increase their economies of volume to make their cars more cost competitive.

The growth in manufacturing capacity of lithium-ion batteries is staggering. According to research from Clean Energy Associates, there was more than 800 gigawatt hours (GWh) of battery manufacturing capacity worldwide in 2020. They project this will quadruple across the decade, exceeding 3,200 GWh of battery production by 2030.

Always audacious, Elon Musk set a goal for Tesla to manufacture 3,000 GWh all by itself by 2030.[43] How much is 3,000 GWh? Assuming an average EV has a 60 kilowatt hour battery, this is enough to build 50 million cars. If US grids were made up of only solar power plants, 3,000 GWh is almost enough storage to power the entire country overnight.[44]

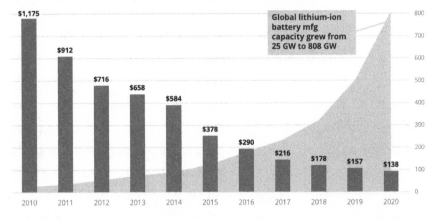

Lithium-ion battery costs decline as manufacturing expands
(US$ costs per kilowatt hour / global manufacturing capacity in gigawatt hours)

FIGURE 4.13 *The costs of lithium-ion batteries have fallen as the capacity of factories has grown. Sources: Clean Energy Associates, RMI, Bloomberg New Energy Finance (freeingenergy.com/g202).*

Batteries share solar's first superpower—they are technologies. Assuming standard-size cells, factories will crank out about 150 billion batteries per year by 2030. This has unleashed colossal economies of volume, driving prices down from $1,175 per kilowatt hour in 2010 to $138 in 2020. Bloomberg New Energy Finance forecasts lithium-ion batteries will drop to $58 by 2030.[45]

To put this in perspective, an MIT study found we can cost competitively replace 95% of all the grid's fossil fuel and nuclear plants with solar and wind when batteries fall to just $150 per kilowatt hour.[46]

Just as consumer electronics drove down the cost of batteries to make EVs affordable, the race for lower cost EV batteries is spawning another huge market—*stationary storage*—residential batteries and grid-scale storage. Low-cost lithium-ion batteries are making home batteries like Tesla's Powerwall increasingly affordable. Will lithium-ion dominate this new market, just as it has in EVs? This is one of the questions I explore in the next chapter.

As with solar panels, batteries can be dual-use. Tesla is building its car frames out of batteries. Other companies are merging solar panels and batteries into a single unit. Batteries take up so little space that they can be placed almost anywhere. Elon Musk once estimated that all the batteries needed to provide 24/7 power to the US could fit in one single square mile of land.

Batteries are also simple. While they sometimes require active cooling systems, the batteries themselves have no moving parts and can be built into racks, just like stereo equipment or computer servers. Most importantly, batteries share solar's fourth superpower: they can easily be located wherever they are being used.

As we will see, batteries and other storage systems are going to be a core part of our future electric system, playing a crucial role at every level of electricity generation, be it a grid-scale wind farm, a community solar project, a solar rooftop, or a tiny system that can power a few LED lights in Africa after the sun goes down.

It is notable that the first part of the battery era freed our devices— mobile phones, cameras, and laptop computers—from constant dependence on the grid. The second part is freeing our transportation from fossil fuels. And the coming era will free our homes and offices from the second-to-second reliance on the grid. Batteries are like bacon. They make everything better.

The biggest battery in the world

Many people are surprised to learn that batteries are more than basic electricity storage—they offer a transformational upgrade to the architecture and economics of electricity. Few examples make this clearer than the changes wrought by one of the world's largest batteries.

For years, South Australia had a power problem. Frequent outages and the need to maximize the value of a large wind farm motivated the government to install a massive battery. Dozens of firms competed for this high visibility project. One of those was Tesla, which makes a grid-scale battery product called Powerpack. In March 2017 Tesla's executives boldly told an Australian newspaper they could install the entire project in 100 days.[47] A prominent Australian entrepreneur tweeted Elon Musk asking if the offer was serious. Musk tweeted back: "Tesla will get the system installed and working 100 days from contract signature or it is free. That serious enough for you?"

The rest is battery history. On November 25, 2017, just 63 days after the contract was signed, the 129 megawatt hour Hornsdale Reserve battery went live, running on Tesla's Powerpacks. At the time it was the largest battery in the world.[48] And almost immediately, it changed everything about Australia's grid. For instance, when a large coal power plant suddenly went offline and threatened to destabilize the region's grid, the Tesla battery responded in 0.14 seconds from 600 miles away and kept the grid from failing until a coal-fired backup plant was able to come online.[49]

The AU$90 million battery's performance exceeded all expectations, saving the region AU$40 million in its first year.[50] "It's regulating the heartbeat of the national electricity market," said Garth Heron, battery owner Neoen SA's head of development for Australia.[51]

Upgrading the grid: from arithmetic to calculus

In less than a year, the Hornsdale Reserve battery erased years of skepticism around the cost, reliability, and impact of batteries on the Big Grid. These lessons are equally true for local energy—batteries are beginning to transform the way we power our homes, buildings, factories, and cities. What makes batteries so special? They remove the most fundamental and intractable limitation of electricity: supply must meet demand, every single second. As engineers might say, the grid is hard-wired into real-time.

Vaclav Smil describes electricity's unique limitation in his book *Energy and Civilization*: "Oil can be stored in tanks; grain in silos; natural gas, in underground caverns. But electricity is the instantaneous commodity; here one second, gone the next. It is a business that operates with virtually no inventory."

We have all experienced what happens when internet services like Netflix get overloaded. Movies will sputter or a site may crash. The grid equivalent is outages and long-term blackouts. Obviously, the grid must be held to a much higher standard of reliability. Occasional spikes in demand should never lead to regular outages. The most fundamental priority in designing the grid is to ensure it can always meet peak demand. This was the subtle brilliance of Edison's original grid design. By aggregating hundreds or thousands of customers into one giant electric circuit, all individual spikes of all those customers are effectively averaged out. The result is a much smoother, more predictable flow that slow-responding coal generators could keep up with.

The fundamental principle of this design is the availability of peak capacity—also called *power*. This is why we collectively refer to the businesses that manage our grids as the *power industry*. Batteries will change the nature of our electricity system entirely. With batteries, electricity can be stored when it is plentiful, like when nuclear plants have excess capacity at night or when the sun is shining during the day, and then reused when it is scarce. This reduces the criticality of peak demand and shifts utility operational focus to electricity delivered over time—which is called *energy*. Batteries are remaking the power industry into an energy industry.

As I will explore in Chapter 6, "Billion Dollar Disruptions," utilities, oil companies, and several other giant industries are all on a collision course for the ultimate control of society's most strategic commodity: energy.

Batteries are freeing electricity from the tyranny of real-time and peak demand. In doing so, they are fundamentally changing the planning, the business model, and the operations of the grid. Edison's original design will need to be completely readdressed. *And for utilities, it will be like shifting from arithmetic to calculus.*

Making electricity smarter and creating the foundation for a next-generation grid

Edison's original elegant grid design—a single giant circuit—is still the fundamental architecture of today's Big Grid. Engineers refer to this as an *analog circuit*, because the grid is managed by small changes in voltage and frequency.

The very first computers were also analog. As innovative as these early computers were, their analog nature made them cumbersome and limited to a single calculation at a time. Vacuum tubes, like the ones in my father's UNIVAC, first allowed computers to switch and route tiny bits of electric power across thousands of small, independent circuits—the *digital* ones and zeros of the modern computer era. This simple yet monumental shift allowed computers to be *programmed* to automatically take multiple steps in sequence, setting them on the path of the world-changing role they play today

Similarly, early telephone systems were analog circuits. When you called someone, the phone company's sophisticated switches would create a direct circuit with the person you were calling. The circuit was analog because the sound waves were represented by small variations in the circuit's power. In the late 1960s, however, telephone switches started their migration to digital circuits. Voices were represented as ones and zeros. Costs declined, quality improved, and entirely new services were created. In fact, it was this transition to digital communications that laid the foundation for the modern internet.

The modern age of computers and digital telephone switches occurred because innovators discovered how to finely control the flow of electricity. This level of control was game-changing, but it was limited to very tiny amounts of power. But a few years ago, this started to change. Scientists found new materials—like silicon carbide and gallium-nitride—that allowed transistors to handle much, much higher power levels.

An entire new class of component was born, called *power electronics*. And with it, entirely new markets and products have emerged. Even if you have not heard of this technology, it is already all around you. Power electronics vary the power into an EV or grid-scale battery so that it charges quickly and safely. Solar inverters rely on power electronics to efficiently transform the direct current in solar panels to the alternating current used on the grid.

Power electronics also make these devices smart. When EVs or inverters are hooked into the grid, power electronics allow them to provide a wide set of *ancillary services* that improve the operation and lower the cost of operating the grid. These include valuable but obscurely named things like frequency control, peak shaving, black start, and voltage support. Think of the grid as a symphony orchestra and each smart inverter or EV as another musician who makes the music just a little bit better.

Power electronics are an essential part of our clean energy future. They are also technologies. As such, they are enjoying growing economies of volume, as did solar panels and batteries before them. You may never actually see a silicon carbide transistor, but know that it is unleashing the full potential of the clean, local energy revolution.

NEGAWATTS

If there is any single person who has most influenced my journey into energy, it is Amory Lovins. I first saw him in 2005 as he gave a

TED talk called "Winning the Oil End Game." He effortlessly wove together stories, data, and witty insights to make the case that the US could wean itself entirely off oil by switching to renewables and investing in efficiency. It was one of the most powerful and positive visions I had ever heard. It planted the seeds for what would eventually grow into *Freeing Energy*.

Many years later, I found myself sitting with Amory in the headquarters of the entrepreneurial nonprofit he co-founded, the Rocky Mountain Institute (RMI). He has an extraordinary mind and an expansive vision. His credentials are too long to list but some highlights include twelve honorary doctorates, a MacArthur Award, and an Ashoka Fellowship. He is also the sole or main author of 750 papers and 31 books, including five foundational works: *Soft Energy Paths*, *Natural Capitalism*, *Brittle Power*, *Small Is Profitable*, and *Reinventing Fire*.

Amory was one of the first people to show that the economics of efficiency and renewable energy could provide what he calls a "soft path" to lower costs and reduced environmental impact. He is a strong proponent that business and capital markets are better options than heavy-handed government policy and expenditures. He was also one of the first people to make the case that small-scale energy systems can be effective, profitable, and resilient. These insights were the first inspiration for my work in clean, local energy.

Early in our conversation, Amory warned that "a lot of computer people want to get into the energy industry, but it can be a real struggle." He said that people from the tech industry, like me, often have little experience dealing with an industry so constrained by policies and reliant on giant capital assets. But he finished his point by commending my efforts to write a book and to learn the industry before I tried to run a business in it. More than anything else, it was that comment that cemented my decision to publish this book.

His insights from that conversation and his lifetime of writings are peppered throughout these chapters, but it was his final point that day in Colorado that I want to share here. After I had taken him through the core ideas I planned to explore in this book, he paused for a moment and told me I was missing one critical idea. He said there was one form of energy that emits no greenhouse gases, creates no pollution, and consumes no fuels. He said the cheapest form of energy is the energy you never use. He was speaking of efficiency, or what he colorfully calls *negawatts*—a typographic error he adopted and popularized.

To really drive this home, I toured his home shortly afterward. It is one of the most energy efficient structures in the world. Finished in 1984, many of its designs have become standard practice around the world. I saw windows that insulated as well as 16 to 22 sheets of glass but looked like two sheets and cost less than three; 99%-passive-solar space- and water-heating; a tight and super-insulated building envelope; and, of course, solar PV on the roof, running his electric meter backward far more than forward. His house even has a full indoor tropical garden, including banana plants.

Not all these systems can be strictly defined as technologies, and many will not reach the billions of units needed to drive the same kinds of economies of volume we have seen in solar, batteries, and power electronics.

But the potential of negawatts can be illustrated by a product that fits squarely into this elite category and has been one of the biggest drivers in the decline in per capita electricity consumption in recent years—LED bulbs.[52] In 2009, LED bulbs cost $25 per kilo-lumen (a measurement of light output). They were a niche specialty product with less than 1% of the market. Over the next ten years, their costs fell 25 times, to $0.90.[53] There were an estimated 10 billion LED bulbs made in 2019 and they made up almost 50% of the worldwide lightbulb market.[54] The International Energy Agency forecasts LED bulbs will achieve 87% market share by 2030, and the US Department of Energy predicts bulb costs will fall to a stunning $0.30 per kilolumen.

LED lighting costs decline as volumes increase
(US$ cost per kilolumen / LEDs as a percent of global unit sales)

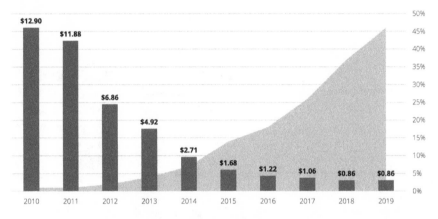

FIGURE 4.14 *The cost of LED lighting has fallen in proportion to the increase in its sales. Sources: IEA 2020 Lighting Report, DOE 2019 Lighting R&D Opportunities (freeingenergy.com/g213).*

The world's love affair with solar was born from environmentalism and growing concerns about the impact of our fossil fuel-powered society. But as we have explored in this chapter, solar has become even more. It is easier to build and maintain, far simpler, and most recently, the cheapest source of electricity ever invented. All of this is due to solar's main superpower: it is a technology, not a fuel.

While many people will continue to struggle with the new economics of economies of volume, solar and its technology sisters—batteries, power electronics, and energy efficiency—will continue their relentless cost declines and feature improvements.

Just how far can these technology costs decline, and what are the implications of super-cheap electricity? Let us take a stroll through the science and business models powering the clean energy industry.

CHAPTER 5

HIDDEN PATTERNS OF INNOVATION

"Seriously, Bill? Have you actually thought this through?" my Uncle Ken asked me.

It was 2016, and I had just spilled the news I had been so excited to share—I was leaving behind my career in technology to jump headfirst into clean energy.

My uncle Ken is very important to me. My first baseball toss was with him. Our family has navigated our share of hardships, and Ken is the person shining the light forward for the rest of us. But at the dinner when I announced my career news, Ken was not enthused.

He went on, "Have you heard of Solyndra?"

"Yes, a little bit," I answered reluctantly. At the time, I knew little about that particular story. All I could recall was that Solyndra's technology worked but the company was a commercial failure. Ken was ready to fill me in.

"The Obama administration sunk $500 million into that solar company, and they lost every cent. This whole solar and renewables thing cannot survive without huge subsidies from the government."

I gulped.

Uncle Ken was not the only person concerned about my career direction. Friends from my days in venture capital were wondering if I had lost my mind. Their reactions to my clean energy career shift usually began with a long slow inhale, followed by something like: "Have you heard about all the venture capital funds that lost their shirts on that space in the early 2010s? Cleantech is really not a good space for people used to high-growth, high-margin businesses."

There were many legitimate reasons to question the future of clean energy back in 2016, and none larger than the massive losses suffered by venture capital firms just a few years earlier, including Solyndra. But I remained optimistic because I had seen similar patterns of early boom-and-bust in other industries and several of them ultimately went on to change the world.

Nonetheless, Uncle Ken's concerns prompted me to take a much deeper look at the recent history of clean energy's boom-and-bust cycle.

THE CLEANPOCALYPSE OF 2011

In 2011, the clean tech industry melted down. Overall, more than $10 billion in venture investments were lost. I call this the *Cleanpocalypse*. The much-heralded potential of clean energy technologies, followed by the bloodbath for investors, spooked venture capitalists and raised questions that continue to color opinions about clean energy today. Uncle Ken's concern for my career is just one example of the fallout triggered by the Cleanpocalypse. And it raised a much larger question: is there something fundamental about cleantech that prevents rapid innovation and makes it incompatible with venture funding? To answer this, let us take a look at the most visible failure of that era.

A 2016 article in MIT's *Energy Initiative* journal titled "Venture capital and cleantech: the wrong model for clean energy innovation," made the case that venture capital, at least back then, was not a fit for clean energy.[1]

Venture capital (VC) firms spent over $25 billion funding clean energy technology (cleantech) start-ups from 2006 to 2011 and lost over half their money; as a result, funding has dried up in the cleantech sector. . . . The results are stark—cleantech offered VCs a dismal risk/return profile, dragged down by companies developing new materials, chemistries, or processes that never achieved manufacturing scale. We conclude that the VC model is broken for the cleantech sector, which suffers especially from a dearth of large corporations willing to invest in innovation.

Solyndra, the cleantech bogeyman

On September 1, 2011, Silicon Valley cleantech darling Solyndra declared bankruptcy. The news was a shock to many because the company seemed to have everything going for it. Solyndra had created a unique, cylindrical design for capturing solar energy that was highly innovative. It had raised more than $1.5 billion, of which $535 million came from a US government loan guarantee.[2] It had even achieved $140 million in revenue in the prior year. Yet somehow, it collapsed. Nearly all the investors, including the US government, lost their money. Overnight, Solyndra became a punching bag for critics of clean energy. It was held up by fossil fuel advocates as proof that renewables were a fad without a commercial future.

Some argued that Solyndra was a rat's nest of political self-dealing. Others argued that the company's technology was never viable, or that the management was incompetent. But none of these truly account for the collapse. The primary reason for Solyndra's failure was far less sensational but no less important. As Varun Sivaram explained in his book *Taming the Sun*, it boiled down to something as simple as ". . . a product that cost $6.29 per watt to manufacture but only earned $3.42 per watt in the marketplace . . ."

Solyndra got out-innovated. So did countless other companies like Nanosolar, Xunlight, eSolar, Konarka, and SolarReserve, that were betting they could find a cheaper way to generate solar power than traditional but expensive silicon-based solar cells.[3]

What happened? With the benefit of hindsight, we can see that there were three unanticipated macro trends that thwarted the ambitious goals of these pioneering innovators.

Cheap polysilicon. The first step in making silicon solar cells begins with wafers, ultra-thin slices of 99.9999% pure silicon. From 2008 to 2011, the price of polysilicon dropped—actually, more like crashed. After reaching a high in early 2008 of $475 per kilogram, it had fallen almost 10 times to $50 by 2011.[4] In three short years, silicon photovoltaics went from being an expensive niche to the cheapest way to generate solar power. It had won the war. Nearly every other approach to capturing solar power, including Solyndra's, went belly up. Polysilicon prices have continued to drop, and by mid-2020, it was $8 per kg.[5]

The natural gas fracking revolution. Natural gas had always been a more environmentally friendly replacement for coal, but it could not compete with coal's low cost. Fracking changed all that. From 2008 to 2012, the price of natural gas plummeted from $12 per million BTU down to $2.[6] Suddenly, renewables were no longer the only option to replace coal. Natural gas plants were not as clean as solar, but they avoided the headaches of intermittency, and they were not limited to sunny or windy geographies. Nathan Myhrvold, the one-time CTO of Microsoft and now co-investor with Bill Gates on nuclear startup TerraPower, told me, "The hardscrabble fracking people did what all of Silicon Valley failed to do—they completely disrupted the energy industry."

The 2008 financial meltdown. The collapse of the mortgage industry brought the sky-high valuations of these cleantech pioneers crashing back to earth and dried up the rivers of cheap venture capital. Even companies whose technologies still held commercial promise were unable to raise additional funds.

There were, of course, other reasons. Some technologies never worked. Some management teams were in over their heads. Others got too focused on scientific breakthroughs and lost sight of commercial realities. A combination of these factors conspired to kill off almost half the startups in the industry.

As we enter the 2020s, cleantech is once again hot. Investors are wiser. Costs are lower. Technologies are more mature. The federal government is more supportive. And, for the first time, clean technologies are cheaper than conventional energy systems. Whether we call it cleantech, energy tech, power tech, or climate tech, we are entering a new chapter in this industry and the potential is now hundreds of times larger.

Of course, the most important success factor needed in clean energy remains the same: making smart, informed choices and building world-class teams of innovators. The big questions are also the same: which technologies and business models will succeed, and which will struggle—and why? In the next two chapters, I will explain the underlying principles of clean energy technologies and introduce you to a few inspiring innovators that are leading the industry into the future.

THE ROOTS OF INNOVATION

One of the most thrilling and memorable events in my research for this book was my dinner with Jim Rogers. Jim, the retired CEO of electric utility giant Duke Energy, is a legend in the power industry, in no small part because he was one of the first big utility CEOs to support the transition to renewable energy. Sadly, Jim passed away in 2018. But I remain grateful for the two hours he spent with me, sharing his story and giving me advice about my new career. It changed the course of my life.

Jim was a student of electric history. His stories of Edison, Tesla, and Westinghouse brought the early industry to life for me and inspired several sections in this book. Jim had recently published his own book, *Lighting the World*, about the tremendous opportunities for electrification among the low-income communities in Africa and India. It is a wonderful book that helped me focus my own passion in this area.

The dinner was long, and Jim gifted me with many great insights. The most powerful was this: "It is time for the power industry to become a technology business again." As a refugee from the tech world, this was music to my ears.

His words were prescient. The power industry is slowly (very slowly) shedding its roots as a fuel-driven, asset heavy, top-down business into something increasingly defined by the economics of technology. This will require from the industry a completely different understanding of economics, because things like solar panels, batteries, electric vehicles, and power electronics all operate on a different business model than fuel extraction, large power plant construction projects, and transmission power lines. The economics of technology are unfamiliar to veterans of the industry, as is the driving force behind technology revolutions—innovation.

Innovation theater

The dictionary defines *innovation* as "The creation of a new device or process that is contrary to established mechanisms." When businesspeople talk about innovation today, they can sound like medieval philosophers talking about alchemy. "Innovation is the secret to creating better products, happier customers, faster growth, and higher profits."

Every electric utility has a section on their website about their innovation. The term has become so popular that its real meaning has been lost. In too many cases, the term innovation is little more than marketing speak—a shiny veneer for an otherwise tired product or project.

But unlike alchemy, innovation is real, and in many businesses, it is alive and thriving, particularly in smaller, fast-growing companies. Under the right conditions, innovation can even transform long-established industries. We are seeing this today with the century-old electricity business, which, despite being one of the largest industries on earth, is experiencing a rebirth built around innovative thinking, driven largely by people from outside the industry.

Cutting-edge technologies, new business models, and tens of thousands of innovators are in the vanguard of this epic transformation. The coming changes will shake up electric monopolies and unleash incredible business opportunities.

I have seen this happen up close before. Several companies I helped lead earlier in my career—in computer networking, the internet, and digital marketing—were part of similar innovation-driven transformations. When I made the leap into clean energy, I devoted myself to learning how my previous experiences with innovation might apply to this industry. I wanted to separate out what critics call *innovation theater* (marketing veneer) from the elusive but very real opportunities for investors and startups.

This chapter and the next are about the underlying patterns of innovation, based upon my experience in other tech-driven industries. With these insights as a foundation, I will forecast the path of the renewable, local energy industry. Whether you are an entrepreneur, policymaker, venture capitalist, legislator, environmentalist, or a concerned citizen, I hope this helps you understand the treacherous divide between wild-goose chases and billion-dollar opportunities.

Patterns from other industries

When I first began thinking about the business of clean energy, I reflected on my experience working as a venture capitalist at a firm called Greylock. In my time there, I met hundreds of entrepreneurs and reviewed thousands of business plans. This gave me an inside view into many of the successful companies the firm had funded. While there were certainly exceptions, I saw three common characteristics that created our biggest returns.

High gross margins. The money left over after the cost of making or selling something is called *gross profit*. When measured as a percentage of revenue, it is called *gross margin*. To illustrate investor economics, I will pick on the humble tomato. If a store buys each one for 80 cents and can sell it for $1, that leaves a 20-cent gross profit or a 20% gross margin. For tech investors, higher gross margins are almost always better. Software companies often exceed 90% gross margins, making them the darling of VC firms.

Asset light. Selling tomatoes is *not* asset light. You need cash to buy them or grow them. You must pay to refrigerate them and, inevitably, you will be stuck with some that do not get sold. In contrast, the digital products of asset light businesses—like music, video games, movies, TV shows, and software—can be duplicated for each customer at virtually no cost, avoiding the overhead of maintaining inventory. This not only lowers costs, but also allows a business to be more flexible and responsive.

Fast growth and disruptive market opportunities. Faster growing companies generate higher returns for investors. Growth allows companies to respond quickly to changing market conditions and to stay one step ahead of competitors. Venture capitalists are not investing in tomato companies because people eat largely the same amount year over year. Tomatoes have no "growth story," as VCs would say. In contrast, software that displaces an existing market and grows rapidly—like Lyft and other ride-share services—disrupts existing industries with something faster, cheaper, and more convenient.

Patterns in clean energy

If software epitomizes fast growth and asset-light business models, the cleantech industry described in the MIT article is the opposite. For example, commercializing a new kind of solar cell is more than just breakthrough science. It requires enormous factories and complex supply chains, both of which consume a lot of cash. This kind of innovation is slow. The journey from lab-bench prototype to full-scale factory can take 3 to 5 years, sometimes much longer. Large customers, like utilities, can take years evaluating a new type of solar cell before they purchase it in volume. And once those solar cells are installed, they cannot be cost-effectively changed or upgraded for decades. No wonder the VC industry ran away from cleantech in 2011. Can the electricity business, or at least parts of it, ever become fast and innovative? The answer is yes.

Like all industries, cleantech is a synthesis of interlinked organizations, technologies, products, professionals, and business models. The journey from raw materials to manufacturing to distribution to customer benefit is complex. Business consultants refer to this as a *value chain.*

I like to think of each big link in the clean energy chain as *orders*, where each order is dependent on the one before it. The first order is *components*, the second is *integrations*, with *services* and *platforms* making up the third and fourth orders. The fifth order, disruptions, is the most exciting of all. For example, in the transportation value chain, tires, steel, and electronics are first-order components that feed the second-order market of automobiles. The third-order market transforms assets into services, like taxis and rental cars. Lyft and other ride services that leverage existing assets are fourth-order markets.

The five orders of cleantech innovation

Disruptions (fifth order) — Shift an existing industry value chain into an entirely new industry

Platforms (fourth order) — Create new markets for existing products and services in return for a commission/fee

Services (third order) — Transform first- and second-order assets into a pay-as-you-go business model

Integrations (second order) — Assemble first-order components into a new product or market

Components (first order) — Small, discrete pieces of more complex value chains

FIGURE 5.1 *The "Five Orders" model illustrates the value chain in clean energy, where each order builds on the one before it.*

While exceptions are common, each of these orders tends toward its own unique set of opportunities, challenges, business models, and financing options. Understanding where a new technology or company fits into these orders simplifies the assessment of its risks and growth opportunities. In the rest of this chapter and the next, we will step through each of the five orders, explaining what they are and offering examples of each.

MANUFACTURING SOLAR AND BATTERIES

Components are the first order in the innovation value chain. They are the foundation for the entire industry. For clean energy, they include solar cells, solar panels, batteries, and power electronics like inverters. The mainstream media loves to herald first-order scientific breakthroughs with headlines like "New Organic Solar Cell Will Reverse Climate Change." These articles are fun to read but usually misleading. First-order businesses are often the most capital-intensive and risky of all types of clean energy companies.

Converting photons to electrons—the quintessential first-order component

More than 90% of solar photovoltaic products are built with cells made of silicon, a material that makes up 28% of the earth's crust. By itself, silicon is rather boring electrically. But if you mix in some chemicals like boron or phosphorous, a silicon semiconductor can suddenly perform electrical gymnastics that are mind-boggling. The nearly 10 billion transistors inside a single iPhone are testament to what semiconductors can do when they are small. When configured for a larger format, silicon semiconductors can efficiently allow incoming photons to dislodge electrons, causing electric current to flow and, voila, you have solar power.

The remaining segment of the solar cell market is based on *thin film* technologies. Whereas silicon wafers are produced in batches, thin films can be made continuously—like newspaper printing. Cadmium telluride (CdTe) is the most popular of this small segment but other chemistries like copper indium gallium selenide (CIGS) have achieved some market share.

If you listen to solar optimists, however, we are just beginning to unlock the possibilities for thin film. A parade of scientific breakthroughs, including quantum dots and organic dyes, has raised anticipation for new ways to convert photons to electrons.

However, the most exciting new thin film PV technology is *perovskites*. In 2009, these cells were limited to 4% efficiency, but the technology has improved faster than anything before it.[7] By 2020, labs were pushing perovskite cells to 25% efficiency, edging into the territory of the best silicon cell efficiencies. Like molecular made-to-order deli sandwiches, perovskites can use a wide variety of minerals within a three-dimensional structure to produce an astounding array of photovoltaic properties. While more than 20 billion silicon solar cells were made in 2020, most of the progress in perovskites remains in labs, with tiny prototype cells. In an irony that only a scientist can truly appreciate, the biggest challenge of perovskite solar cells is that they break down in the presence of, well, sunlight and humid outdoor air. Undeterred, scientists, investors, and governments are investing hundreds of millions of dollars into perovskites, all toward offering better, cheaper solar power.

Batteries: a chemistry lab in your pocket

Solar technology is far from simple, and scientists and entrepreneurs who seek to improve it have many avenues they can explore. Cost per watt, efficiency, degradation over time, and environmental impact: each can be a focus for innovation. As complex as designing and building solar cells is, batteries are even more complicated. Batteries add more characteristics, including charge rate, cycle count, charge depth, self-discharge rate, energy density, and thermal stability, to name just a few.

Dr. Jeff Chamberlain is a globally respected scientist and the CEO of battery-focused venture capital firm Volta Energy Technologies. Prior to funding battery innovations, he was helping create them as the head of commercialization efforts for the top US battery research center at the time, Argonne National Labs. Jeff explained the unique complexity of batteries to me in this way: "When your phone is charging or discharging," he said, "matter is literally being moved around inside that battery. It is an ongoing chemical reaction happening in your pocket."[8]

As of 2020, lithium-ion is the most popular chemistry for the batteries powering the transition to clean energy. Just as silicon solar took an early industry lead thanks to the existing semiconductor value chain, lithium-ion has an early advantage in storage thanks to a decade of powering laptops and mobile phones. Because it is lightweight and energy dense, lithium-ion is ideally suited for these mobile applications as well as the growing market for electric vehicles.

All lithium-ion battery cells have three main parts: an anode, a cathode, and an electrolyte. Charging the battery frees up electrons in the anode, which is usually made of a carbon material like graphite. The electrons travel through an electrolyte made of lithium salts and settle into the cathode, which is commonly made with an oxide of nickel, manganese, and cobalt (NMC) or its close cousin, nickel-manganese-aluminum (NMA). Discharging the battery drives electrons the other way. Another popular chemistry, lithium iron phosphate (LiFePo) weighs more but supports more charge cycles and is more thermally stable, squelching lithium-ion's propensity for catching on fire. This added safety makes it a growing favorite for *stationary storage*—using batteries to balance out the intermittency of solar in residential and grid-scale projects.

In 2016, energy consultancy Bloomberg New Energy Finance (BNEF) predicted lithium-ion batteries would fall from $1,000 per kilowatt hour in 2010 to $120 by 2031. In 2020, as prices fell below $150, BNEF revised its forecast down to $58 per kilowatt hour by 2030.[9] Like solar, the innovation in batteries is moving faster than even its biggest supporters can predict.

All this progress is based on the same basic architecture John Goodenough helped commercialize decades ago. A radical new approach is finally making it out of the lab that could change the industry. *Solid-state* batteries replace the liquid electrolyte with a solid ceramic material. Solid-state batteries promise big leaps in charge cycles, thermal stability (no fires), and energy density. Even battery hero John Goodenough is getting into solid-state. In 2014, he said, "I want to solve the problem before I throw my chips in. I'm only 92. I still have time to go." So, two years later, at the young age of 94, he announced a breakthrough for his own glass-electrolyte solid-state battery. In 2020, at age 98, he announced commercial partnerships to bring it to market.

All these innovations, and the billions of dollars fueling them, are just the first part of the value chain—the first-order components. As I will show you later, storing electricity becomes even more innovative and exciting when we move past small components into multi-component integrations, opening up an entirely new class of energy storage. But first, I want to look at why it is so hard to transform these laboratory innovations into commercial successes.

Making stuff is hard

The next time you read a breathless article about some new clean technology that will singlehandedly make fossil fuels obsolete, be sure to digest it with a healthy dose of skepticism. I love reading these articles as much as anybody, but the reality is the vast majority of these scientific breakthroughs will never leave the labs and most of the funded component companies will fail. The industry refers to the challenge of commercializing prototype products as "crossing the valley of death."

Why do so many promising scientific breakthroughs fail to achieve commercial success? It turns out that it is extremely hard to make large volumes of high-quality, low-cost products. Prototypes can be handcrafted by experts who obsess over every detail. Each part or raw material can be carefully vetted before it is applied. If a problem emerges, the prototype can be reworked or revised until it is perfect.

Mass producing products requires an entirely different set of processes and skills from making prototypes. Automation needs to replace delicate human interaction and requires that each step be performed flawlessly the first time, every time. Raw materials arrive in bulk and every bit must meet stringent specifications. Setting up a finely tuned system like this requires time, money, and highly specialized expertise. Commenting on this in late 2020, Elon Musk tweeted: "The extreme difficulty of scaling production of new technology is not well understood. It's 1,000% to 10,000% harder than making a few prototypes. The machine that makes the machine is vastly harder than the machine itself."

A hare becomes a tortoise

Investing in first-order innovations is a brutal game that can trip up even the brightest business minds. Take, for example, Bill Gates and the famed VC firm Kleiner Perkins, part of an investor group that provided almost $200 million to Aquion, a startup with a novel battery built with a salt-water chemistry.[10] In 2010, it set an audacious cost target of $300 per kilowatt hour, less than one-third of lithium-ion batteries at the time. By 2016, it had achieved its cost goal. But like all the solar companies in the Cleanpocalypse that lost the cost race with cheap silicon, plain old lithium-ion batteries had already become cheaper. In the summer of 2017, the company shut down.

Crossing the valley of death

Across solar, battery, and other energy tech industries, a clear pattern emerges. First-order markets trend toward standardized, commoditized products. Buyers love this because components are typically interchangeable, allowing them to source from multiple vendors and negotiate better prices. But for vendors, the pressure to reduce prices is crushing. This leads to companies building bigger factories with more automation, so that they can take advantage of economies of volume—the main reason solar and battery technologies have plunged.

Huge economies of volume are game-changing, but building massive factories creates an entirely new problem—it takes much longer for them to break even. This effectively locks vendors into manufacturing the initial product much longer. During that window, it is often too expensive and risky to try anything new. This strategic conundrum affects entire industries, and it explains a big reason why silicon solar cells and lithium batteries have remained dominant components, even when potentially better innovations are now available.

Just ask a computer professional why the internet is stuck using an outdated protocol like TCP/IP, its core mechanism for transferring data, and you will hear a very similar story. Or, as Mark Dudzinski, the former Chief Marketing and Strategy Officer for General Electric so aptly explained to me, "the early leaders usually win the race."

If economies of volume lock entire industries into legacy technologies, how can innovations like new solar cells or battery chemistries ever make it to market? The answer to this billion-dollar question is debated in boardrooms, consulting firms, and government offices across the world. Here are three example strategies that show it is quite possible to break out of this strategic paradox.

Making existing value chains better. This is the most common strategy, and it is exactly what the smartest perovskite companies are doing. Rather than inventing new manufacturing processes and technology standards, perovskite innovators are applying their breakthroughs to improve existing silicon cells.

Creating a second electricity-producing layer makes these *tandem* or *multi-junction* cells even more efficient. It is helpful to think about incremental innovations with some example numbers. Consider an early innovation with a $50 million market. A 1% improvement creates $500,000 of value, probably not enough to cover the cost of bringing the innovation to market. But in a mature market of say, $1 billion, a 1% improvement creates $10 million in value—a far more attractive opportunity for any aspiring innovators and investors. Multiply this by dozens of 1% innovations and you can see how large, mature value chains manage to continuously improve and why brand-new innovations struggle to ever catch up.

Being far, far cheaper. Most breakthroughs focus on price, but this is a far riskier strategy than it appears to be. In fact, most of the failed companies from the 2011 Cleanpocalypse went belly-up trying to create cheaper alternatives to early silicon solar cells. So rather than beat the mainstream market leader, smart companies can pick a small but growing adjacent market and focus on lowering costs for that emerging niche.

For example, in 2020, companies making a new type of storage, *long-duration storage,* were betting that buyers storing electricity for weeks and months would accept fewer recharging cycles in exchange for much lower costs. Limited charging cycles would not work for frequent, short-term applications like EVs, but long-duration storage could be a breakthrough for homes, allowing them to store power for a week of rainy weather.

Create new functionality. While price always plays a predominant role, new functionality is the most exciting way for new first-order innovations to break into existing markets. Lead acid, nickel–metal hydride (NiMH) and nickel–cadmium (NiCd) rechargeable battery technologies were the early leaders in batteries for electronic devices. But lithium-ion batteries have largely eclipsed these older chemistries despite being more expensive. It turns out that new features like lighter weight and more forgiving charging cycles can command a higher price and still take market share. Companies creating solid state batteries are looking to turn the tables and do the same to traditional lithium-ion. They believe that huge improvements in safety, weight, and charge speed will be enough to motivate buyers to pay a premium price. Watch this space carefully.

Once a new chemistry, a new product, or a new architecture crosses the "valley of death" and attains the status of an industry standard, few things are more important than low price. And few places on earth are better at producing massive volumes of low-price products than China.

CHINA'S EDGE

Large scale mass production is rocket fuel for first-order technology businesses. About 70% of the world's solar cells and panels are made in China and Taiwan. To understand China's path to global dominance, I traveled to Shanghai to meet with one of the world's foremost experts on Chinese solar and battery supply chains, Andy Klump, the CEO of Clean Energy Associates (CEA).

Andy is a global citizen. He grew up in St. Louis, the son of a grade school teacher and a taxicab driver. As a teenager, he created a lawncare business that helped him pay his own way through private school. Today, he lives in Shanghai, where he leads CEA's global team. He stands out in the county not just because of his six-foot five-inch height, but because he speaks fluent Mandarin and navigates the country like a native. He is both earnest and constantly in motion, like a blend of Jimmy Stewart's character in *Mr. Smith Goes to Washington* and Aragorn the Ranger from *The Lord of the Rings*. (Full disclosure: Andy and I both work with each other's businesses.)

Clean Energy Associates is the leading solar and battery supply chain advisory, quality assurance, and engineering services firm. Andy describes his company's unique value: "The global supply chain for solar and batteries is a finely tuned instrument. It is impossible to overstate the costs, delays, and even safety issues that can emerge from even small product defects or delivery hiccups. We work with our clients to make sure every single product they buy arrives exactly as promised and at precisely the agreed upon time and location. We ensure their supply chains are a competitive advantage."

There are several reasons China became so dominant in solar manufacturing, several of which defy the sound-bite answers often thrown around. Many people assume China's secret is low-cost labor. That may have been true decades ago but today, labor rates have risen and places like Mexico are actually lower cost.[11] Besides, increasing levels of factory automation are making labor costs less material to the final price of panels. For example, labor accounted for only 3%–5% of total costs in 2018.[12] In fact, one cell and module factory I visited was entirely automated. The only people moving around were tending to the machines.

Another common view is that Chinese solar manufacturers are financially propped up by their government. Although this is true, China is not the only government supporting its solar industry. The US federal government, along with county and state governing bodies, have provided hundreds of millions of dollars in financial support for many early solar manufacturers. And as one Chinese executive argued to me, most of their solar companies are traded on US stock exchanges—so anyone can see their financials and that the government's support is not as large as many claim.

So if we cannot point purely to labor and subsidies to explain China's success, what else is going on? To help me get a broader understanding of China's unique advantages, Andy and I visited the headquarters of one of the world's largest solar companies, Jinko Solar. There, we met Gener Miao, Jinko's Chief Marketing Officer.

As we sat in Gener's office, he shared what he believed were China's two biggest competitive advantages. Years ago, he told us, the Chinese government decided that solar manufacturing was strategic to its national interests, and set out to become the world leader. The government's approach was to set up the entire supply chain, from raw materials to finished products, in a single geographic region. Whenever Jinko runs low on a raw material, he explained, he picks up his phone, and a few hours later a truck arrives with all the materials he needs. If this happened anywhere else in the world, he said, Jinko would have to go through a complex procurement process that might take days or weeks.

Gener explained this supply chain advantage is essential for the second big reason behind China's success: "You need to keep your factories running continuously. Every hour of every day, all year long. Every minute your factory sits idle due to equipment issues or lack of materials means you are less competitive and less profitable." He said every solar manufacturer in the world understands this, but he believes Chinese companies are world leaders in factory uptime.

China has emerged as the solar manufacturing center of the world, but other regions, like Europe and India, are not content to let this stand. Motivations like national energy security, domestic jobs, and, frankly, national pride are all at stake. It is anyone's bet whether China stays on top.

(The global tug-of-war that resulted in China's ascension to global solar leadership is fascinating. If you are interested in learning more, I recommend Greg Nemet's deeply researched book, *How Solar Energy Became Cheap,* or listen to my podcast with him for an abbreviated version.)

Components through the decades and into the next

"At the turn of the century, we didn't know which energy tech would win. Now that argument is largely done," states cleantech pioneer Danny Kennedy, the Chief Energy Officer at the New Energy Nexus.[13]

By 2010, silicon solar and lithium-ion batteries had established themselves as the component technologies to beat. From there, the crux of the competition shifted from science labs to manufacturing plants and China has set a pace for everyone else to follow.

Global leadership in battery manufacturing is still up for grabs, and China is working hard to replicate its dominance in solar. But private companies like Tesla and governments like the European Union are wary of ceding another energy technology, so they are making huge bets to build their own capacity in hopes of a more balanced global marketplace.

For me, the bigger strategic question is not which region will ultimately lead in component manufacturing, but rather, does it even matter. Every step up in orders leads to the potential for more differentiation, higher profits, and faster-paced innovation.

Let us turn our sights to second-order innovations.

FAR MORE THAN THE SUM OF THE PARTS

Second-order innovations integrate a variety of first-order components to create a new product or market. Moving up the value chain brings three big benefits, all of which come a lot closer to the sweet spot for venture investors and entrepreneurs. Returning to our example of food, if vegetables are first-order products, then soups and salads are second-order products.

Here are the most important characteristics of second-order businesses:

Higher profits. Even a limited number of vegetables can be combined into a nearly infinite variety of soups. This diversity counterbalances commoditization and price pressure and boosts the profitability of second-order products.

Increased speed. Growing vegetables on a farm or building a solar factory are slow processes. But making soup or installing solar panels on a roof are much faster. Second-order products can more quickly be prototyped, tested, and iterated, resulting in faster learning curves than their first-order cousins.

Less upfront capital. Buying a bushel of tomatoes is cheaper than planting a large farm, and a pallet of solar cells is far cheaper than building a factory. Lower capital requirements attract a much wider set of investors *and* allow a higher number of integration-focused startups to enter the market.

How do second-order businesses work in energy tech? The earliest and most prolific second-order businesses in the clean energy industry were solar installers. They procure components and assemble and integrate them into working systems. Because each customer, each field, or each rooftop is unique, each installation is customized. It is nearly impossible to automate and streamline the entire process. As the cost of solar declined, customer demand increased and installation firms proliferated. But differentiation was slim and so were profits. Nonetheless, the industry took off. From 2010 to 2020, there were more than 6,000 firms installing local rooftop solar. Most of these firms are small, local businesses, with each averaging just 21 employees.[14]

Solar, solar everywhere

The rapid rise of solar installers was just the opening act of an epic play. The next act is faster-paced and has attracted a much wider set of innovators and entrepreneurs. This new generation of second-order businesses are moving beyond labor-heavy business models and creating entirely new products, some of which can be mass produced in automated factories.

Prior to 2020, nearly all solar projects were built the same way: flat panels sitting in fields or on rooftops. But the marvel of photovoltaics is that many surfaces exposed to sunlight can be redesigned to also convert photons to electrons. Do a Google search on solar windows, solar paint, solar walls, or solar driveways, and you will see that the future of solar is not constrained to black rectangular panels.

Collectively, these innovations hint at a growing revolution in local energy called *building integrated photovoltaics* (BIPV). For most of these, like transparent solar windows or solar facades that look like wood or stone, widespread commercialization is still years away. For the near term, the big action is on roofs, most of which sit directly in the sun all day long.

Consider a traditional rooftop solar installation. A specialized installation team starts by analyzing the roof to ensure it can support the weight of panels. They then climb onto the roof, secure rows of metal racks, and screw down panels. If the roof leaks or shingles need to be replaced, repairs are cumbersome. All this labor is in addition to the original roof construction.

But what if the roof or shingles generated solar power directly, and no add-on panels were needed? Labor and design would be cut in half. Most of these new products will be much smaller than traditional solar panels. But the key point is, regardless of their size, they combine the function of generating solar energy and protecting a house from the weather. This emerging industry is poised to revolutionize local energy. Best of all, most of the early innovators are making their solar roof products using off-the-shelf, standard solar cells, benefitting from the value already created by first-order economies of volume. These purely second-order BIPV companies can move swiftly. They can prototype, test, and learn from new ideas in weeks and months, rather than the years it might take to commercialize a new first-order solar cell technology.

The newest generation of rooftop BIPV is called *aesthetic solar*. It uses cutting-edge materials that make traditional cells look like asphalt shingles or terracotta tiles. Tesla was the first big-name company to offer this type of solar roof. While Tesla's "shingles" are more expensive than traditional panels, Elon Musk described the benefits at the product launch in 2016: ". . . a roof that looks better than normal roofs, generates electricity, lasts longer, has better insulation, and has an installed cost that is less than a normal roof plus the cost of electricity."

Much like the way Apple and Steve Jobs delighted the world by making computers that were both functional and attractive, BIPV creates entirely new ways for innovators to differentiate their products and delight customers.

A 2021 study out of Germany estimates that a combination of rooftop and BIPV facades can provide 40% of the electricity consumed by a standard-size office building. If local battery storage is added, these solar technologies can provide almost 60% of the building's annual power consumption.[15]

The promise of solar buildings is tantalizing, but for their costs to follow the declines of traditional panels, they will need to tap into economies of volume and build a new generation of factories and automation.

Making the machines that make BIPV

The future of low-cost BIPV is being pioneered in the picturesque Italian town of Padua, about an hour west of the famed canals of Venice. Nestled among terra-cotta roofed homes sits a small industrial park and the headquarters of Ecoprogetti, one of the world's most innovative solar manufacturing equipment makers. The company's name literally means "eco projects" in Italian. To see the bleeding edge of BIPV, I traveled to Ecoprogetti to meet Laura Sartore, the company's charismatic and animated CEO.

Passionate and self-deprecating, Laura speaks quickly, leaning in to emphasize important points. Fortunately for me, English is one of the many languages she speaks fluently. She laughed as she told me about her father, the company's founder: "So being a real Italian, his first PV panels were cooked and baked inside the oven in our family's kitchen." Even before the company embraced BIPV, Laura explained that local energy was in their DNA. "Our mission is to allow people from any part of the world to start new production of photovoltaic panels locally."

FIGURE 5.2 *Bill Nussey, author, and Laura Sartore, CEO of Ecoprogetti, on the BIPV roof of her headquarters. Credit: Bill Nussey.*

The company's own buildings are a masterwork of BIPV. Two of the four-story walls have curved facades built with hundreds of gorgeous interspersed solar panels. Laura described her headquarters building to me: ". . . the roof is built entirely of solar panels. They are built with four- to six-millimeter-thick temperate solar glass, and they are encapsulated with an insulating material." Her roof is a giant solar panel, and it generates 180 kilowatts, enough to power her entire operation and still have excess to sell back to the grid.

FIGURE 5.3 *The decorative BIPV panels on the exterior façade of the Ecoprogetti headquarters building in Padua, Italy. Photo courtesy of Ecoprogetti.*

Building the machines that automate panel manufacturing is a rare competency. Her equipment needs to nimbly move big pieces of panel glass that can easily weigh 40 pounds. It then precisely and delicately positions each solar cell. Finally, all this is wired together with hundreds of paper-thin conductive ribbons. The assembled panels are then "cooked" at strict temperatures and tested with ultra-accurate sunlight simulators. It is like bulldozing and brain surgery all wrapped into one. Manufacturing BPIV panels is even tougher because they require unique materials and specialized processing.

BIPV is still an early market, but Europe is leading the way. In areas like Padua, with a rich architectural history, traditional solar runs afoul of heritage preservation laws. Laura asks rhetorically, "Can you imagine these beautiful red terra-cotta roofs covered with black solar panels?" Unsurprisingly, this is one of the markets Ecoprogetti's equipment is transforming. Some of her customers make terra-cotta-colored panels that blend imperceptibly into Italy's famous red roofs.

Innovating solar racks

The first time I attended a solar trade show, there were more than 100 companies selling racks—yes, those dull metal structures that secure solar panels to rooftops and fields. It was tempting to underestimate the value of the structural products. To my untrained eye, it looked like an endless parade of gray steel beams with the occasional motor or gear (one company painted their racks green, but I cannot really call that an innovation). Even though expensive steel and high winds make designing competitive racks tricky, the racks were largely interchangeable, competing on narrow features and price. But a new generation of innovative new designs is taking local energy to entirely new places.

As I discussed in the last chapter, there are so many dual-use locations for solar that we could practically power the entire US. Parking lots, roads, ponds, lakes, dams, farmland, and, of course, the rooftops of homes and buildings are all opportunities to generate solar power. What do they all have in common? They all need innovative ways to safely and cost-effectively mount the solar panels, wires, and power electronics. Racking may look like dull steel beams, but the second-order innovations here will make the difference on whether thousands of projects get built or are written off as too expensive.

To Lithium and beyond

All these solar innovations are exciting. But by themselves, they are like trying to ride a bike with one pedal. If we want to see local energy take off, we need a second pedal, and that pedal is storage. But will storage ever be cheap and plentiful enough to balance out the incredible growth of solar?

Most of us do not give our freezers a second thought. But if you went back to 1920 and told someone they could inexpensively keep their food fresh for months, they would look at you with a mix of skepticism and awe. Similarly, in the 2040s, we will be entirely blasé about the terawatt hours of storage that let us use daytime solar around-the-clock, even if it seems inconceivably expensive and advanced in the early 2020s.

The first-order battery storage I discussed earlier is a big start. For lightweight, frequently recharged applications, lithium-ion is out in front of its competitors. Experts agree that $100 per kilowatt hour is the tipping point where EVs become cost competitive with gasoline cars. Some vendors claim to have already reached that price. However, storage will have to be even cheaper than this to power your home or community during a week of rainy days. Is it possible to go even lower?

Second-order electric storage solutions are built by integrating different components into a single system. As I said earlier, there are a nearly infinite variety of second-order "salads," so the potential mechanisms for storing electricity are as mind-bogglingly diverse as they are creative. I will touch on three examples of different types of next-generation storage and focus on integrations that are relevant for local energy, both community scale and smaller. I have no doubt that this level of diversity and innovation will lead to electric storage being as ubiquitous as your kitchen freezer.

Flow batteries. These turn traditional battery cells inside out. Rather than locking the electrolyte inside a tiny, fixed-sized metal cylinder or pouch, flow batteries separate electrolytes into a pair of liquids that are stored in much larger external containers. In normal batteries, increasing the kilowatt hours to be stored requires a linear increase in the number of battery cells. In flow batteries, more kilowatt hours just require larger tanks. Flow batteries are complex machines, so only a few pioneering manufacturers are trying to make them small enough for residential use. But for larger, commercial- and community-scale projects, flow batteries may become a serious competitor to lithium-ion.

The most common flow batteries use a chemistry built on vanadium, a rare and expensive material. But other chemistries being commer- cialized include iron, saltwater, and zinc. Even if flow batteries prove to be more expensive than lithium-ion on a kilowatt hour basis, their designs allow for a nearly unlimited number of charge cycles, mak- ing them ultimately less expensive to operate in many applications.

Thermal storage. Traditional batteries convert electric energy into chemical energy and back again. Thermal storage replaces chemical reactions with temperature as the intermediary energy store. Malta, a company that originated inside Google, uses excess electricity to heat up a proprietary salt and later uses that stored heat to gener- ate electricity. Another company, Highview Power, is pioneering the opposite approach. It super-cools plain old air into a liquid for storage. To generate electricity, the air is heated in order to expand and turn a generator. These systems are novel and exciting, but the laws of thermodynamics make it difficult to get round-trip charging cycles as efficient as chemical batteries.

For example, if you put 10 kilowatt hours of electricity into a storage system, lithium-ion might give 90% back, or 9 kilowatt hours. Even with the best thermal storage systems, you might get only 5 or 6 kilowatt hours back. The cost of thermal storage will have to drop considerably to make up for its lower efficiencies.

Mechanical energy storage. In addition to chemical bonds and temperature extremes, it is also possible to store electricity mechani- cally. These second-order systems take advantage of gravity, momen- tum, or pressure. In fact, the oldest and largest grid-scale storage in use is mechanical. It is called pumped hydro.

It starts with a hydropower dam, which produces electricity as water flowing to lower elevations turns a generator. Pumped hydro simply uses electric pumps to push water back up to the reservoir behind the dam, effectively "recharging" the system. The US has 23 gigawatts of pumped hydro, most of which was built in the 1970s.[16]

But a host of new, more effective mechanical storage systems are being developed. *Flywheels* use electric motors to spin heavy discs, and those same motors can be used as generators to convert the spinning motion back to electricity. The discs usually require exotic materials (to keep from flying apart at high speeds) and need to be inside airtight containers to reduce friction, making this system too expensive for all but niche applications.

Compressed air storage uses electric pumps to move pressurized air underneath water bodies, into containers, or even into underground caverns and old oil wells. Releasing the air spins turbines that generate electricity. Unfortunately, in many cases, compressing air generates unwanted heat and lowers the roundtrip efficiency below that of chemical batteries. A Canadian company called Hydrostor has a proprietary system that captures and reuses the excess heat, increasing its overall efficiency. Hydrostor has several large storage projects in operation, with many more in development.

It is hard to predict which types of non-battery storage will find the biggest commercial markets, but it is a good bet that many solutions will be entirely outside the box, borrowing from components and technologies that are outside the traditional energy value chain. This brings me to one of the most radical and promising storage systems, and one of my personal favorites.

Forty stories of rocks

In the early days of the internet, Bill Gross created one of the most famous startup incubators, IdeaLab. I first met Bill at an internet conference in 1997 where he hosted a "cocktail napkin business plan competition"—each submission had to fit on a single napkin. To my surprise, I won the competition with an idea for a web application that scanned the internet for resumes and made it easy for recruiters to quickly find candidates, kind of a proto-LinkedIn. My prize was a job offer with IdeaLab. For better or worse, I was happily employed at VC firm Greylock, so I was not able to accept the offer. Nevertheless, Bill became an entrepreneurial hero of mine, and his vision for new ventures has only become more exciting over time.

Bill was one of the first internet entrepreneurs to make the move into clean energy. He has since launched a handful of cleantech companies, several of which have blockbuster potential. When I made a similar leap several years later, I reconnected with Bill. I wanted to learn all about one of my favorite examples of second-order storage innovations: Energy Vault.

Energy Vault is a mechanical storage solution that uses a 400-foot-tall, multi-headed crane powered by renewable energy, to lift enormous blocks made of cement and sand, which it stacks in towers. When the sun is shining or the wind is blowing, the crane draws in electricity to lift and stack blocks on top of each other. In their current design, the stack can reach over 40 stories high, and the tower has several independently moving cranes to increase the rate the system can be "charged." When the sun is not shining or the wind has stopped blowing, the system can discharge, by taking a block off the stack and letting gravity slowly pull it downward. The motor that lifted the block is then used as a generator, sending electricity back into the grid.

I asked Bill how they came up with such a radical out-of-the-box approach. His answer exemplifies the iterative process of technology innovation:

Energy Vault is the result of a long series of trials and failures to try to make low-cost energy storage. I looked at the lowest-cost energy storage in the world—pumped hydro—and tried to figure out a way to beat that, but in a scalable way that doesn't require the topography (a mountain with a reservoir on top and on bottom for the water to reside).

I abstracted out the essence of pumped hydro—gravity storage where you lift a weight up to store energy and then lower it to get that energy back—and tried to figure out other mechanical ways to replicate or improve upon that at scale.

I tried ski lifts carrying buckets of gravel up a hill. I tried large silos filled with rocks or water. I tried conveyers building up mounds of dirt. After many unsuccessful tries, we ended up with the idea of stacking blocks to make a tower. The reason that worked so well economically was that the blocks provided both the weight AND the height.

We have now proven the round-trip efficiency is higher than pumped hydro, and we have customers around the world, but we have yet to scale this anywhere near the potential global demand. It's now a race that we can't wait to participate in.

Each Energy Vault system sits on roughly two acres of land and can store 20, 35, or 80 megawatt hours of electricity—enough to meet the daily needs of 600 to 2,500 homes. At scale, it is targeting a cost of $150 per kilowatt hour and a round-trip efficiency of 85%. This is about the same as lithium ion in 2020, but Energy Vault will have a nearly infinite number of charge/recharge cycles, making its long-term cost less expensive.

This system is too large for a home, but it is the ideal solution for a small community or office park. The beauty of Energy Vault is that it is environmentally positive—its primary component is rocks. It is an ideal way to dispose of concrete and other construction debris. Nearly all its components are widely available in competitive marketplaces, which keeps component costs very low. And similar to flow batteries, the total amount of storage can be cost-effectively expanded just by stacking blocks in a wider circle.

The transformational power of integrating components

Most of these second-order electric storage systems are effectively big machines that make more sense for larger local energy projects like campuses, neighborhoods, and towns. But what about smaller projects, like homes and small buildings? First-order batteries are likely to remain the preferred approach because of their sheer simplicity and their cost-effectiveness at small scales. But even here, the opportunities for second-order integrations are transformative, even if they are less headline grabbing.

I am going to share two stories that make this case, both of which highlight Tesla. To be clear, I do not advocate putting the company on a throne or fawning over its CEO, Elon Musk. In fact, I believe some of the company's tactics are questionable. That being said, critics too easily dismiss Tesla's success as mere marketing and hype. What makes Tesla a valuable object lesson—here and in other examples in the book—is the company's extraordinary product and engineering skills. Because some of its breakthroughs are more second-order than first, these breakthroughs can seem obvious in hindsight. Let me start with Tesla's residential storage product.

The first version of Tesla's Powerwall was essentially a first-order component—a standalone battery pack. Like so many competitive products, it required a specialized third-party solar inverter to coordinate the power flows between the house, the grid, and the solar panels. This multi-component system made sense because the parts could be put together any way the buyer desired. It offered maximum flexibility and choice. Unfortunately, it also required highly trained electricians and installers to design and configure each installation.

When the upgrade shipped a few years later, Tesla's Powerwall 2 did something few competitors had tried before. It integrated a custom inverter into the same chassis as the battery pack, effectively elevating it from a first-order product to a second-order product. The Powerwall 2 initially faced skepticism because the built-in battery inverter duplicated some of the functionality and costs already in the third-party solar inverter.

This may seem like a small, mildly provocative change, but the product designers at Telsa knew exactly what they were doing. The built-in inverter proved to be a game-changer. The integration of all these components transformed the installation process. Now, the design requirements and expertise were so straightforward that an electrician with the know-how to install an electric stove or dryer could install it. (Tesla, nonetheless, requires training and certification for its installers.) Because Powerwall 2 was now set up via a phone app, the installers did not have to know how to configure digital controllers. By mid-2021, Tesla had already shipped 200,000 Powerwalls.[17]

In 2021, Tesla demonstrated yet another benefit of an integrated system and did something no one could have imagined: the company increased the peak and continuous power output of *existing* Powerwall 2 products by 50%, simply by pushing a software update.

Tesla's Powerwall makes a remarkably strong case that second-order integrations can lower installation costs, reduce complexity, and streamline ongoing system management, all at the same time.

Setting the stage for even more innovation

If you really want to appreciate the power of second-order integrations, consider another Tesla product, its groundbreaking Model S electric sedan. The car had few, if any, proprietary components. Every electric car before it incorporated specialized, proprietary batteries. Tesla's early cars used widely available laptop batteries. The car's control electronics were built with off-the-shelf parts. In short, GM, Ford, BMW, and every other auto maker had access to the same components that Tesla used to build the Model S. Yet Tesla managed to integrate those same first-order components into what *MotorTrend* magazine crowned the "Ultimate Car."[18]

> The Model S changed the way the world thinks not only about electric cars but also about cars in general . . . it remains clear there isn't another vehicle created during our 70 years of existence that has had a truly comparable effect on automobiles, the automotive industry, and society at large.

At the most fundamental level, Tesla's only advantage—and arguably the most important one of all—was its exceptional engineering and design, both second-order integration skills. By the time you read this, Tesla may have flamed out or, as I hope, continued its amazing product innovation. Either way, the company is once-in-a-generation, and it is risky to draw conclusions about any patterns of cleantech successes from it. Fortunately, the opportunities for innovation in cleantech are limitless and most do not require singlehandedly disrupting a trillion-dollar global industry.

The first decade of the 2000s was a battle between several first- and second-order value chains. After the dust settled, wind turbines and silicon solar PV were the clear winners. The 2010s saw an enormous rollout of clean energy. The economies of volume drove down costs, making local energy projects increasingly cost competitive with electricity from the Big Grid. As first-order markets matured and standards were locked in, they became a foundation, the bottom level of a pyramid that makes possible the next level in the value chain, second-order innovations.

The 2020s are already seeing exciting new businesses and investment opportunities. At every step along this evolution, clean, local energy just keeps getting cheaper and more reliable for end customers. Climbing higher in the value chain speeds the pace of innovation and creates even more opportunities for entrepreneurs and innovators. In fact, this is where it starts to get really exciting.

CHAPTER 6

BILLION DOLLAR DISRUPTIONS

FIGURE 6.1 *Orville Wright watches as his brother, Wilbur, takes history's first powered flight on December 17, 1903. Photo courtesy of the US Library of Congress.*

Every American school kid learns the story of Orville and Wilbur Wright. On December 17, 1903, just outside Kitty Hawk, North Carolina, these brothers created history by achieving the world's first controlled, powered flight.[1] After centuries of failed efforts, this accomplishment was widely believed to be impossible. Simon Newcomb, a widely respected officer of the National Academy of Sciences and the director of the American Nautical Almanac Office, summed up the expert opinion of the time: "No possible combination of known substances, known forms of machinery, and known forms of force can be united in a practicable machine by which men shall fly long distances through the air."[2] It may be one of the worst predictions in history, especially since it was made three years after the Wright brothers first flew. The news of their flight, which had received little national press attention, had apparently not yet reached Newcomb.

The Wright brothers were able to build their pioneering aircraft by combining several innovations for the first time. They used wind tunnels to run tests. They designed a sophisticated three-axis control system. But there was one particular innovation that allowed these two brothers, whose main experience was running a bicycle shop, to defy scientific experts like Newcomb.

To understand the genesis of this innovation, we need to go back seven years and consider another set of innovations in an entirely different industry. On November 16, 1896, the largest source of electricity in the world at the time—the Niagara Falls hydropower plant—was switched on.[3] It energized a newly built grid, lighting up homes and industry all the way to Buffalo, 22 miles away. This engineering masterpiece was the brainchild of Nikola Tesla and his business partner, George Westinghouse. Never before had there been so much electricity available so affordably.

How do these two stories intersect and how did their respective years of innovation come together to give birth to commercial aviation? The answer is aluminum.

Cheap electricity did more than light homes. As Gretchen Bakke explains in her book *The Grid*, "It was in Buffalo with the force of Niagara's own power, where the manufacture of aluminum first became a profitable enterprise." Extracting aluminum from ore requires extraordinary amounts of energy. Before the Niagara plant slashed the cost of electricity, aluminum was as rare as gold. Within a few years, this extremely useful, lightweight, corrosion-resistant material was disrupting dozens of industries.

It is easy to understand Newcomb's skepticism. He, like many others, had done the math. He knew that the best gasoline engines of the day weighed too much to power an aircraft. The Wright brothers knew this, too, because they had already reached out to the major engine manufacturers and learned that none had an engine that could meet their power and weight specifications. To the experts, the lack of a light engine meant powered flight was impossible. To the Wright brothers, the lack of an engine meant they had to build their own. They turned to their shop mechanic, Charlie Taylor, and asked him to make one. Charlie did something that was unheard of back then—he built the engine block with cast aluminum, which had only recently become affordable. The rest, as they say, is history.

The unexpected intersection between a hydropower dam and powered human flight highlights the often unexpected nature of disruptions. They rarely occur in straight lines. Experts frequently miss what is right in front of them. It is often outsiders like entrepreneurs and innovators that connect the dots and unleash the disruptions that reshape entire industries and society.

We have already discussed components and integrations. Now let us explore the ways these building blocks can be assembled into higher-order innovations that are even more interesting and competitive. We will start with the third order of innovations.

NEW POSSIBILITIES FOR POWER

The third order in the cleantech value chain is *services*. These build on existing assets from first- and second-order markets and transform their business model, most commonly turning these assets into a service. For instance, if vegetables are first-order and soups are second-order, then restaurants are third-order. For other industries, third-order business models could include taxis, internet applications, air travel, and mobile phone calls.

When it comes to third-order businesses, the Big Grid may be the most famous and impactful of all. As I have said earlier, Edison's first grid was certainly a technical triumph, but it was the brand-new business model of sharing centralized electric infrastructure across many customers that truly changed the world. Before Edison's centralized Pearl Street Station, electric generating systems were so large and so expensive that only the ultra-wealthy could afford them.

Electricity became even more affordable and available in the following decades as Insull pioneered the regulated monopoly business model. This not only made large, centralized generation even larger but also locked out any competition that might threaten profit streams. Insull's business model has worked so well, it remains largely unchanged a century later.

The hidden costs of success

Power plants have gotten bigger. Power lines are longer. Electricity has become reliable, affordable, safe, and universally available (in most places in the world). But as we know now, this success has come with an array of hidden costs that are growing larger and spilling out into the lives of everyday people. Pollution and greenhouse gases are damaging the planet. The environmental impact of mining, drilling, and disposing of wastes are harming communities. And despite cheaper technologies like solar, wind, and batteries, the electric bills we all pay are steadily increasing.

How are utilities responding to these growing issues? Most of them are doing only what the law requires of them to resolve these issues while at the same time doing everything they can to slow the coming changes. It is a form of weaponized inertia. And as electricity consumption per capita peaked in 2005 and is now declining, some of the utilities' guaranteed profits are shifting from building new power plants to funding lobbying designed to lock in their monopolies more deeply.[4]

Sadly, this is the reason why my section on third-order businesses in today's electricity industry is so short. Electricity services are a monopoly. Only one company in each region in the US (and most of the world) can sell a kilowatt hour. With no competition, innovation is stifled and often nonexistent. It is deeply ironic to me that inspired innovators like Edison and Tesla would no longer have a place in the very system they built.

Local energy offers customers a choice for the first time

For the first time, small local energy systems like rooftop solar and microgrids can actually be cheaper than the Big Grid. For the first time, third-order electricity markets can be created outside the laws protecting monopoly electric utilities.

How does this work? In most parts of the world, monopoly laws grant a single company the right to *sell* a kilowatt hour of electricity. Competition is shut down before it can get started. But those same laws say little about *generating* a kilowatt hour for yourself. Insull and the early architects of electric monopolies never anticipated that a solar "power plant" on a roof could be *cheaper* than the vast economies of scale offered by their centralized grid. Producing your own electricity falls outside the monopoly laws protecting utilities. This is the driving force behind rooftop solar and the early momentum of local energy.

It does not stop there. Several states, including California, Georgia, and Colorado have created carve-outs that *specifically* allow third parties to sell kilowatt hours, as long as *the equipment is installed on the same property where the kilowatt hour will be consumed.* This exception has given rise to a billion-dollar industry and companies like Sunrun, Sunnova, and Solar City.

These early innovators and dozens of others like them have shifted the traditional business model that required an upfront installation fee to something more like a subscription model, in which the company owns the hardware and charges homeowners for each kilowatt hour. It is like an electric utility, but cheaper. For homeowners, these companies look a lot like a utility except that electricity is generated locally. This third-order business model gives customers direct control over the reliability and environmental impact of their electricity, all while they are saving money.

As the price of first-order components continue to decline, local energy installations will continue to grow. Lower costs are allowing batteries to be added into a growing portion of these installations, creating systems that are less and less reliant on Insull's grid business model. As second-order markets like solar installations and storage systems continue to mature, the number of third-order electricity providers will accelerate. For the first time in history, utilities have real competition. And as I will explain in the next chapter, they are not taking this lying down.

CHOREOGRAPHING ELECTRONS

Fourth-order businesses increase the value of existing services, assets, and products by bringing them to new markets or adding new functionality. Businesses become *platforms* when they make money helping other organizations conduct business.

The fourth-order business model is as old as commerce itself. Flea markets and consignment shops work as fourth-order businesses because they charge merchants a commission for providing access to a big, new group of prospective customers. Extending my food metaphor, if vegetables are first-order, soup is second-order, and restaurants are third-order, then delivery services are fourth-order.

The growth of online shopping, digitized commerce, and the rise of online marketplaces like eBay and Etsy have taken the platform business model to new levels. Apple's App Store takes a cut every time a software application is purchased through its platform. In the software world, Salesforce.com is one of the most well-executed platform models, because entire applications are built on top of it, each of which creates a long-term revenue stream for Salesforce. Fourth-order businesses like Uber, Lyft, and Airbnb, create entirely new revenue opportunities for people who own cars and homes and previously had no way to monetize their assets. (Sometimes these businesses become so successful, they spill into and disrupt unrelated industries. When this happens, they qualify as fifth-order companies. More on that shortly.)

Being a platform business is the most coveted position in any industry, not just clean energy. This type of company is catnip for venture capital investors. Unlike business models earlier in the value chain, fourth-order companies are not encumbered by capital assets, so they are able to move relatively quickly.

Creating new fourth-order markets is risky, but when they are successful, they are often very profitable. Third-order businesses often balk at sharing their revenue with fourth-order platforms, but access to more customers or new revenue streams can be compelling. Of course, this can come at a cost: Postmates and Uber Eats may expand the number of customers a restaurant reaches, but in addition to taking a cut of the restaurant's profits, they also eliminate what is special about the in-person eating experience, like service and ambience.

Platforms for energy tech

How do platforms (fourth-order businesses) work in what energy tech veteran Danny Kennedy describes as "choreographing electrons." It is like moving from a world with a few giant corporations that turn a few large knobs to one where millions of individuals—aided by sophisticated algorithms—all turn their own small knobs. Large utilities are reluctant to use platforms run by outside companies, but local energy opens entirely new markets for fourth-order platforms to coordinate and control these small, independent systems.

The heart of these new platforms is their software, which the industry broadly refers to as *Distributed Energy Resource Management Systems* (DERMS). Traditionally, grid managers like utilities and wholesale market operators depend on a variety of extremely technical services to ensure uninterrupted grid operation. These include frequency regulation (to stabilize AC power flows), peak shaving (smoothing consumption to reduce expensive demand spikes), demand response (reducing or shifting consumption at the customer's site), and operating reserves (space capacity set aside to handle unexpected changes in generation or demand). DERMS use software to coordinate small portions of hundreds or thousands of local energy systems to create grid-sized solutions that are lower cost than traditional infrastructure. Most of these emerging innovations are only possible because of digital technologies. Let us look at a few approaches to choreographing electrons in the new world of local energy.

Smart homes and buildings. The cost for utilities to generate our electricity varies widely across the day. Most of us pay a fixed rate that averages out our utility's costs, so we give little thought to when we run our washing machines or air conditioners. It turns out that coordinating our *loads* with utility generation costs can save both parties a lot of money. Intelligent or *smart* systems fine-tune when our heaviest electric loads run so they are optimized to utilities' costs. Smart thermostats like Google's Nest are an early success story, reducing electricity bills and making homes more comfortable at the same time.

Electric water heaters are a large, untapped opportunity. As long as the water temperature is fairly constant, homeowners could care less when the heating coils are turned on. Coordinating the precise timing of these loads across thousands of electric water heaters could dramatically cut some of the most expensive peak generation for utilities. The fourth-order business opportunity is to coordinate the heating coil loads and split the savings between utilities and homeowners.

Many commercial buildings pay a kilowatt hour rate that varies by their peak usage. *Demand-side management* (DSM) can coordinate heavy loads like air conditioning, water heaters, dryers, pumps, or even factory equipment to ensure they do not all turn on at once, driving up the peak rate. *Demand Response* (DR) takes this idea one step further. Rather than optimizing to utility pricing, utilities actually pay customers to reduce their demand. This is a win-win proposition.

When peak demand gets particularly high, utilities have traditionally been forced to ramp up their most expensive power plants called *peaker plants*. New digital communication solutions are providing an alternative that is often cheaper. Utilities can now ask some customers to reduce their peak demand by shifting the use of their air conditioners or water heaters to a later time. Balancing out peak demand with DR saves so much money that utilities can pay their customers with the money they save.

As buildings add their own solar and battery systems, even more options emerge to reduce the peak power drawn from the grid and thus lower electricity rates and lower bills.

Virtual Power Plants (VPPs). A traditional local energy system is too small to matter to the Big Grid and the utilities that operate it. But if you coordinate thousands of these smaller systems, their combined output can approach the output of a traditional centralized power plant. A VPP operated by Sunrun in California is helping offset the cost of solar and battery systems for 500 low-income homes by tapping into a small portion of each battery's capacity and aggregating all the power to act like a small peaker plant. Each of the 500 families gets paid for sharing their resources, and the utility has extra power available to help it replace a retiring fossil fuel power plant.[5] This innovative platform disproves the common utility talking point that local energy systems are disruptive to the Big Grid. In the future, small local energy systems will be an integral (and lower cost) part of providing electricity to everyone, not just the local energy owners.

Transactive markets. Today, all electricity is bought and sold through a single entity, the utility. Monopoly laws make any other approach illegal. But as consumer pressure mounts for the monopolies to relax these rules, fourth-order platforms will begin to allow individual local energy owners to buy and sell *from each other,* becoming what the industry calls *prosumers.* Like an electric version of a local farmers' market, these systems will manage the complex physics and accounting of electricity flowing between hundreds—or millions—of individual systems. In the last chapter, I will present the opportunities, challenges, and inevitability of those markets.

Microgrid controllers. In Chapter 3, "The Rise of Local Energy," I shared the story of the Stone Edge microgrid and the amazing technology from a company called Heila that coordinated the project. Technologies like Heila's not only provide a platform to coordinate every part of a microgrid and reduce costs for local energy systems, but that same intelligence can integrate with the Big Grid, simultaneously offering demand management, smart load services, and transactive markets.

Fourth-order opportunities in the clean energy industry are in their infancy. There are still many years for ambitious market leaders to emerge. This type of business requires few assets and limited capital, so innovation can be more rapid than with its first-, second-, and third-order counterparts. The unique characteristics of fourth-order businesses will attract a great deal of venture capital, further fueling the growth of this part of the value chain.

Utilities do not need to be on the losing end of this wave of innovation. They still own the massive distribution systems essential for many of these business models to work. In fact, if the utilities are willing and able, they can profit from the biggest fourth-order opportunity in the industry, what industry strategist Peter Fox-Penner calls *smart integrators*. This business model parallels several of the ideas listed above by transforming utilities from sellers of electrons to choreographers of electrons. Such a move would undoubtably increase their profits and future-proof their business models.[6] However, as I will discuss in the next chapter, few utilities are likely to do this voluntarily.

BIG OIL VS THE BIG GRID

When Uber and Lyft launched on Apple and Android phones, they were just two more apps among thousands available on the mobile internet marketplace. But to the taxi industry, these humble apps— downloadable in a few seconds—represented a magnitude 9 earthquake. What makes this disruptive is that Apple, Uber, and Lyft did not have to work with taxi companies or get their permission. They sidestepped the entire existing car-for-hire industry merely by pushing a small app onto our smartphones.

When an outside value chain crosses into an unrelated but well-established industry, incumbents are almost never prepared.

If the change happens quickly, as is usually the case with technology-driven value chains, the incumbents simply cannot adapt fast enough. This is what I call a fifth-order innovation, or what business thinkers call a *disruption*. No discussion of disruptions is complete without quoting from the definitive book on this subject (and my personal favorite book), *The Innovator's Dilemma*, by Clayton Christensen.

> When [disruptive technologies] first appear, they almost always offer lower performance in terms of the attributes that mainstream customers care about . . . But disruptive technologies have other attributes that a few fringe (generally new) customers value. They are typically cheaper, smaller, simpler, and frequently more convenient to use. Therefore, they open new markets. Further, because with experience and sufficient investment, the developers of disruptive technologies will always improve their products' performance, they eventually are able to take over the older markets.

Ten years ago, solar was the disruptive technology—it was more expensive than fossil fuel-powered electricity, but it was cleaner and simpler. Now solar is the cheapest way to generate electricity, and the new disruptor is local energy.

Think of the disruptions you have seen in other industries. Internet sites decimated newspapers; FedEx's overnight shipping and electronic mail replaced many US Post Office services. A century ago, gas lighting was replaced by electric lights. In each case, existing market structures were upended. Enormous new companies emerged as incumbents became less relevant and even faded away. Customers enjoyed more desirable products and services. Underlying all this change is a massive and rapid shift of profits from incumbents to disruptors. This is nirvana for investors like VCs.

Let me share a brief case study on one of the most visible disruptions of our day. In New York City, from 2015 to early 2020 (before the pandemic), Uber and Lyft grew from 60,000 to 750,000 rides per day—a 12 times increase.[7] During the same time, taxi rides fell by half, from 400,000 to 200,000. These new services did not simply replace taxi rides, they expanded the entire market, nearly doubling the number of rides taxis had performed. At the same time, in just five years, ridesharing siphoned billions in global taxi revenues into the pockets of independent car owners around the world, creating employment for hundreds of thousands of people. Riders enjoyed a far more convenient *and* cheaper transportation solution (not to mention tidier cars that were better maintained). And, of course, Uber and Lyft took a small fee with each ride. Ridesharing represents one of the biggest returns in VC history.

To be entirely fair, the impact of the rideshare industry is complex and its path to success was littered with unsavory actions, but there are nonetheless many valuable lessons to be learned as we map out disruptions in the power industry. For instance, taxis are effectively a regulated monopoly, just like the power industry. Prices are set by government regulators and only a select few organizations are granted the right to offer taxi service, effectively curtailing competition. Also, like the electric utilities, the taxi industry used a range of legal and legislative efforts to prevent the onslaught of competition.

In the end, at least for most cities, strong consumer support for a better and cheaper product effectively overrode the taxi monopolies. Competition emerged, prices declined, services improved, and customers were the winners. I have little doubt the power industry will follow the same path.

Electricity disrupting oil and gas

Disruptions are coming at the existing power industry from every direction, and it understandably feels threatened. But the power industry has its own once-in-a-lifetime opportunity to use its inherent advantages to create a fifth-order disruption against *another* industry—the mammoth oil and gas business.

Consider two other markets that are transitioning from fuels to electricity. First, about half of the $160 billion annual US natural gas production is used for heating, both for comfort and industrial processes.[8] Much of this heating will eventually be electric. Second, US automobile gasoline is a $500 billion market that will eventually be replaced by electricity.[9] Whether it takes one decade or five, both these fossil fuel value chains will be replaced by cheaper, safer, and more reliable electricity. *Electrification* is at the heart of the transition to clean energy, and it represents an enormous opportunity to double the size of the $400 billion US electricity industry.[10]

These transitions are classic fifth-order disruptions. The increased opportunity to sell electricity is business as usual to the power industry, but the loss of revenue to oil and gas companies is an industry-wide disruption. Of course, within the electricity industry, the allocation of this revenue windfall between big utilities and local energy is very much up for grabs.

Electric monopolies: the disruptor or the disrupted?

The prize is a trillion dollars of annual energy spend. And that is just in the US. The electric monopolies have deep legal protections and trillions of dollars of infrastructure in place. They are in the best position to seize these opportunities and solidify their market dominance for another century. But as I will explore in Chapter 7, "Utilities vs the Future," they will almost certainly fall short.

The oil and gas companies are not sitting idly by waiting to be disrupted. While they will hold on to their profitable fossil fuel sales as long as they can, most realize that the future is electricity, and their eye is on the prize. These companies bring financial scale that dwarfs electric utilities. For example, Royal Dutch Shell and BP are each ten times larger than utilities like Southern Company, Duke Energy, or PG&E.

Oil is also one of the most ruthlessly competitive markets on earth. In contrast, electric monopolies have spent a century with government-guaranteed profits, and not a whiff of competition. These two mega-industries are on a collision course—Big Oil is stepping into the Bgi Grid. A top executive at oil giant Shell told an audience of energy executives in 2019: "The amount of power—of clean power—we will need to be selling . . . will make us by far the biggest power company in the world. Many [electric utilities] are at a disadvantage, because they have this enormous legacy position, with coal plants and nuclear plants, but also a very centralized philosophy." In other words, if the oil and gas companies have their way, competing with utilities will be like an 800-pound gorilla versus ceramic garden gnomes.

As electric utilities battle oil giants, they are facing another assault in their own backyard. Local energy is growing faster, getting cheaper, and is a better fit for the values of a 21st century world. But all is not lost for the electric incumbents. If they can shake off old habits, reinvent their business models, and embrace the risk of innovation rather than shunning it, there is a clear, almost obvious path forward, which I will get to later in the book.

Regardless of how the market evolves, two things are certain. The electricity market will soon restart its once rapid growth and the cost of generating electricity will plummet. The only question is whether the electric monopolies will remain a part of this new energy economy or, by failing to reinvent their business models, they will get cut out entirely.

As utilities navigate their way through all these changes, several massive disruptive electric opportunities are emerging. I will step through three of these next: fuels made with electricity, super-abundant electricity, and using local energy technologies to electrify the billion people living in energy poverty.

The first of these, which I call *Fuels 2.0*, represents the biggest opportunity for the Big Grid to become the disruptor, rather than the disrupted, and to take on one of the most powerful industries in the world, Big Oil.

FUELS 2.0: HYDROGEN

Electrification is threatening to disrupt the oil and gas industry. Electric vehicles will soon be cheaper than their gasoline predecessors. Electricity is slowly replacing natural gas for heating and industrial processes. But as I said earlier, the oil and gas industry is nothing if not ferociously competitive. Even as it invests to sell electricity, it will not easily give up on fuels. Capturing carbon emissions from fossil fuels is the most obvious bet, but it inevitably drives up prices and does not solve broader environmental problems like particulate pollution and wastewater from drilling.

So the industry is making another enormous bet. This one is based on a new kind of fuel—one that is just as clean and green as solar and batteries. Burning it creates no greenhouse gases and no pollution of any kind. This new fuel allows the oil and gas industry to maintain all the competitive advantages it enjoys from its tremendous distribution and storage assets as well as its well-honed global business model. This magical-sounding fuel is hydrogen (H2), and you may be surprised to learn that it is already a mature and well-established market, generating $119 billion in 2019 in the US alone.[11]

Hydrogen: the good, the bad, and the ugly

Hydrogen power is not a new idea. Back in 1875, Jules Verne imagined its potential as a fuel in his novel *The Mysterious Island*. One of his characters tells his companions that, in the future, ". . . water [will be] decomposed into its primitive elements, and decomposed doubtless, by electricity, which will then have become a powerful and manageable force . . . Yes, my friends, I believe that water will one day be employed as fuel, that hydrogen and oxygen which constitute it, used singly or together, will furnish an inexhaustible source of heat and light, of an intensity of which coal is not capable."

In many ways, hydrogen is a perfect fuel. When measured by weight, it is almost triple the energy density of natural gas.[12] Thanks to the process of electrolysis, it can be made with nothing more than electricity and water. When it is burned for heat or used in a fuel cell to generate electricity, its only byproduct is water.

But hydrogen has its downsides. It is extremely volatile: ten-times easier to ignite than gasoline.[13] Images of the infamous 1937 *Hindenburg* disaster, when a German dirigible burst into flames, remain etched in many people's memories. Hydrogen has other complications. Like natural gas, it is very hard to compress, so pipelines are the most cost-effective mechanism to distribute it. Unlike natural gas, however, hydrogen is corrosive to standard metal pipes and containers, so its ability to piggyback existing infrastructure is limited.

Perhaps the biggest drawback to any hydrogen-based energy system is that nearly all the hydrogen in use today is made from natural gas, called *grey hydrogen*. Even though hydrogen itself is clean, the process of making it has all the downsides associated with fossil fuels. But that is changing. Renewable electricity, particularly the low-cost excess solar electricity that I described earlier, is making electrolyzed hydrogen, called *green hydrogen*, increasingly cost competitive and arguably the cleanest fuel ever created. This process is broadly referred to as *power-to-gas*, or P2G.

Bill Gross, whom I introduced in the last chapter with the mechanical storage company, Energy Vault, has started another clean energy company, called Heliogen, that offers an innovative new approach for creating green hydrogen. His company has raised more than $500 million from Bill Gates's Breakthrough Energy Ventures, the US Department of Energy, and an IPO.[14] Using artificial intelligence, Heliogen's technology precisely focuses reflected sunlight to generate ultra-high heat, which can be used in industrial processes including the production of green hydrogen.

Hydrogen can make fossil fuel companies look green

Hydrogen represents a rare point of agreement between the fossil fuel industry and clean energy advocates. Fatih Birol, the Executive Director of the International Energy Agency (IEA), says, "I have rarely seen, if ever, any technology that enjoys so much political backing around the world. Countries who have completely different views on energy and climate all [jointly] saying that hydrogen is a key clean energy technology."[15]

As pressures from environmental and climate advocates mount, green hydrogen is the only fuel considered as environmentally friendly as solar and wind. Oil and gas companies are investing billions in it because it fits comfortably with their fuel-based business models, and it provides them a legitimate and immediate story about going green. Green hydrogen may be the best option for these companies to retain their market power in the coming decades.

Darius Snieckus, the Editor-in-Chief at renewable energy media company Recharge, describes hydrogen as ". . . both a fig leaf and a drag chute for oil and gas companies." Many experts are skeptical that green hydrogen can ever become cheap enough to compete with grey hydrogen, not to mention renewables like solar and wind. I think this analysis misses the point. The question is not about the cost to produce and distribute hydrogen. Instead, the question is how much it is worth to the oil and gas industry to have a bridge that keeps them relevant and powerful as we move into a world powered by clean energy. I believe this is a much larger number and, as such, hydrogen will play a much larger role than it would on its costs alone.

Hydrogen: the disruptor

Green hydrogen is a fifth-order disruption, transforming industries that are historically unrelated. While oil and gas companies are among hydrogen's most vocal supporters, the technology to generate, store, and distribute it pulls in a much wider set of organizations including governments, science labs, and countless startups and venture investors.

Even though lithium-ion batteries appear to have won the race against hydrogen for powering the short driving distances of passenger vehicles, hydrogen may well prove superior for long-haul transportation like trucking, railroads, and ocean-going shipping. Several groups are also exploring hydrogen for aviation, because it packs more energy per pound than jet fuel and far more than batteries. Sweden is even piloting steel production, using hydrogen as a replacement for CO2-heavy coal.[16]

Somewhat ironically, about half of the US hydrogen production is used to refine oil. This segment will inevitably shrink in the coming years.[17] The other half of hydrogen production is used to manufacture ammonia, the key ingredient in fertilizer.[18] Ammonia can also be used as a fuel to generate electricity, with some experts predicting it will play a larger role than hydrogen in a post-hydrocarbon energy future. Ammonia is less likely to ignite and much easier to compress and transport. While ammonia has its own challenges, like being toxic in concentrated form, its use in farming means it is already safely distributed all over the world.

Hydrogen as a battery

Electrical storage is the most exciting application for hydrogen and ammonia. The industry calls this *power-to-gas-to-power* (P2G2P, really). Like flow batteries, adding more storage to a hydrogen "battery" simply requires adding more containers. This makes hydrogen a promising solution for long-term electricity storage. For local energy, hydrogen could be used to augment or replace batteries and single-handedly solve the problem of week-long rainy or cloudy weather, when too little solar is created.

For hydrogen to reach its potential, two things need to happen. First, major innovations are required to bring down the costs of its first-order components like electrolyzers, fuel cells, compressors, and storage containers. Next, the second-order systems built from these must become more efficient. Currently, the round-trip efficiency of a hydrogen "battery" is often below 50%, meaning half the electricity going into the system is lost during compression and conversion. Billions of dollars of research are being poured into every aspect of this value chain and there are already dozens of high visibility *green hydrogen* pilot projects across the world. Most of the near-term opportunities for entrepreneurs and investors are in first- and second-order businesses but there is a lot of room for differentiation and profitable companies across all five orders.

YOUR EV IS A WIRELESS GRID

Electric vehicles are casting utilities as the disruptors of the much larger oil industry. The market for transportation energy is entirely up for grabs as automobiles shift from internal combustion engines

(ICE) to electric vehicles. A 2019 report by consulting firm McKinsey says, "EVs will become the lower cost option [compared to ICE vehicles] in the coming 5-10 years."[19] Taking that idea further, Bloomberg New Energy Finance predicts half of all cars sold in 2040 will be EVs.[20] This represents a huge growth opportunity for electric utilities.

But the value of EVs to electric monopolies goes well beyond increasing revenue. Most automobiles are driven only about an hour a day; the rest of the time they sit idle.[21] Suburbanites' cars typically sit in a garage connected to a charger. Most EV owners pay the same price for a kilowatt hour throughout the day, so they tend to charge their car at the end of the day. This is creating an ever-larger demand peak that will eventually exceed local capacity for some parts of the grid. But with dynamic pricing or even simpler programs, EV owners can be incentivized to charge in the middle of the night, when utilities have abundant power, and their generation costs are the lowest.

An even larger opportunity for collaboration between EV owners and utilities is called *Vehicle-to-Grid* (V2G). Rather than merely shifting charge times, properly equipped EVs can be called upon to discharge their batteries back onto the grid, allowing utilities to tap them as a replacement for peaker plants. (This is the epitome of a fourth-order business model.)

In May 2021, Ford rocked the automotive world when it announced Lightning, a fully electric version of the iconic F-150 pickup truck. Among the truck's many impressive features, it was the first mainstream electric vehicle to tout built-in V2G. According to Ford, the Lightning can power a home for three days! Volkswagen has joined the convoy by announcing that all its vehicles will be V2G-enabled in 2022.[22] With Ford and VW boldly embracing V2G, customers will soon see this as a required feature, just like power steering.

One of the biggest concerns about V2G is that powering a home will increase the number of recharge cycles on the EV's battery, potentially reducing the lifetime of the vehicle. Fortunately, both Tesla and General Motors have recently announced plans for "million-mile batteries" that could last for decades, removing any concerns that V2G will prematurely wear out an EV.[23] We can expect most other EV manufacturers to follow suit in supporting V2G and long lifetime batteries in the coming years.

The potential storage from car batteries is enormous. Gerbrand Ceder, a professor at the University of California, told CleanTechnica: "The amount of stored energy traveling around on four wheels is much greater than any utility company will ever build and put on the grid." A 2019 study from Lawrence Berkley National Labs looked at this idea for the state of California.[24] They found that scheduling the electric vehicles to charge in the middle of the day, rather than the evening when people get home from work, would be like adding 1 gigawatt of storage and would save the state $1.45 billion to $1.75 billion in up-front stationary battery costs. V2G was even larger. They found that if just 30% of workplace chargers and 60% of home chargers allowed EVs to provide power to the grid, it could offset up to $15.4 billion in grid-scale storage.

V2G promises to be a huge opportunity for utilities. But it is also a Trojan Horse for local energy. In a twist on conventional thinking, V2G makes grid defection and true off-grid living a real possibility. How will this work? In what is called *vehicle-to-home* (V2H), your EV's excess battery energy can flow back over its charging cable into your home or your home's battery, rather than the grid.

On average, Americans drive about 30 miles per day, which takes about 10 kilowatt hours of their EV's battery storage.[25] This leaves a lot of battery headroom for V2H, especially since most days your solar panels and home battery will be able to provide all the power your house consumes. EVs will come in handy when a string of rainy days occurs and solar output drops. During these periods your car can automatically charge its batteries a bit higher at the office or the local charging station. When your car plugs back into your house, it sends this surplus electricity to your home battery, effectively making up for the lower levels of solar.

This idea becomes even more disruptive when V2H extends beyond your personal vehicle. In the future, every electrified Amazon delivery truck, Uber and Lyft car, and pizza delivery vehicle could be a source of recharging for your home's battery during rainy weeks. Entire fleets of electric vehicles could be dedicated solely to refilling home batteries. These vehicles could all easily be fitted with oversized batteries that allow them to charge multiple homes along their routes. This is just a new variation of the well-established truck network that distributes heating oil or refills large propane tanks.

Because the industry loves acronyms, I have decided to call this mobile-battery-to-home or MB2H. The roadside assistance vehicles that recharge drained EVs on the side of highways are already practicing this business model at a small scale. MB2H is one of the largest fourth-order business opportunities in local energy and it will undoubtably be a fifth-order disruption to the electric monopolies and their power lines. With apologies to Doc Brown, the loveable scientist in *Back to the Future*, "Where we are going, we won't need powerlines."

The largest impediment to this vision is the legality of MB2H selling electricity by the kilowatt hour to your EV in your garage (or directly into your house). Ten years ago, this might have been a transaction that only utilities could have provided. But incredibly, thanks to policymakers' excitement to see EV charging networks built out, 32 states have already exempted EV charging stations from utility monopolies in some form or another, allowing third parties (companies that are not monopoly utilities) to bill customers by the kilowatt hour for charging their cars.[26] The need for this exemption is fairly clear and likely to become universal. Prior to these exemptions, EV chargers billed customers for the duration of their recharge—like paying for gas by how long you keep the nozzle in your tank rather than by the gallon.

If all this seems fantastical, consider this: twenty years ago, it would have been considered unthinkable to disconnect your home's telephone landline. Even today, many people complain about the quality of their mobile phone's connections at home. Yet 50% of Americans have already "cut the cord" on their telephone landlines, and the trend continues.[27] While the reasons for disconnecting from the grid are not the same as for adopting mobile phones, the point is that consumers are more willing to make once-unthinkable changes when it comes to convenience and cost. As I said earlier, I do not expect mass adoption of these grid alternatives to happen anytime soon. But their existence fundamentally shifts the power we have as electricity customers and local energy pioneers.

SUPER-CHEAP EXCESS ELECTRICITY

Since the Big Grid was rolled out to most of the US in the 1940s, the price of electricity has been remarkably consistent and stable. Solar is on the verge of changing this, driving prices far lower than anything ever seen in energy history and triggering another of the big disruptions I present in this chapter. There are two emerging factors that will accelerate the cost declines in solar even more dramatically.

First, most experts are coming around to the realization that the problems posed by the intermittency of solar and wind can be partially addressed by *overbuilding* these projects. As I will explain in Chapter 8, "The Battle for Public Opinion," it is almost always cheaper to build an additional megawatt hour of solar than it is to buy a megawatt hour of batteries to store it. As clean energy guru Michael Liebreich told me, "Overcapacity is not a bug, it's a feature. If you have 2 cents per kWh solar, then even if a third of it is surplus, you're left with 3 cents per kWh power—that's still half the price of power from any conventional source."[28]

Only time will tell how far this strategy can go, but it is certain that there will be a lot of excess solar power, particularly in the longer days of summer. This is already impacting markets like California, where the state literally pays neighboring Arizona to take its excess electricity generated on sunny summer days. This odd financial arrangement will be short-lived as innovators find ways to capitalize on those cheap electrons.

Second, because solar requires no fuel, its costs eventually become negligible. In 20 years, much of the 100 gigawatts of US solar power plants will be paid off.[29] After that, any electricity they generate is *practically free*! Lawrence Berkeley labs reports the lifetime of solar plants continues to lengthen—from 21.5 years in 2007 to 32.5 years in 2019.[30] This is enough to power almost 20 million US homes at virtually no cost for a decade or longer.[31]

Aviation was an early and unexpected beneficiary of cheap electricity when it transformed aluminum from a precious metal into an inexpensive commodity. What will become possible when solar makes electricity an order of magnitude cheaper? I could devote an entire book to the innovations and investment opportunities that will emerge when summertime electricity plunges in cost. I will offer only a short list here.

Creating fresh water. A global crisis is growing as many parts of the world are running short on fresh water. Desalinization technology, which removes the salt from sea water to make it drinkable, requires a tremendous amount of electricity and is currently too expensive for all but the most water-parched areas. Cheap solar electricity will cut out one of the costliest parts of desalinization and help bring it mainstream. Another technology, still early in its development, uses dehumidifier-like technology to literally pull water from the air. This is also electricity intensive, but it can work anywhere in the world, even in desert-like climates.

Indoor agriculture. Inexpensive electricity will also help us address crises in food production, particularly in regions experiencing extended droughts or flooding. Cheap electricity can help power facilities that grow crops indoors, especially at the times of the day and year they require the most electricity.

Carbon capture. Experts agree that even dramatic cuts in greenhouse gas emissions will not be enough to achieve international goals. Technologies like *direct air carbon capture* remove CO_2 from the atmosphere, but they generally require large amounts of electricity. This market is still in its infancy, but it will necessarily become massive. Since greenhouse gases are present in the air year-round, the cost of removing them can be lowered by using cheap summer solar. Beyond capturing carbon, innovative companies are finding ways to turn the extracted carbon and waste plastics into useful products like new plastics and fuels. These conversions consume a lot of energy as well, so cheap electricity will make them commercially viable much sooner.

Data centers. Our computers and phones are increasingly dependent on cloud computing. The vast data centers that we connect to use an estimated 1% of global electricity, and that is expected to grow.[32] In 2020, Google announced it was testing *load shifting* low priority computation tasks to run when excess low-cost solar and wind electricity are available.

Hydrogen. As I will explain shortly, hydrogen holds tremendous promise as a replacement for natural gas, because it can use the existing fossil fuel distribution infrastructure. Cheap, excess solar accelerates its path toward cost competitiveness.

Fertilizer. Plants require nitrogen to grow. Most fertilizers contain this essential nutrient in the form of ammonia or NH3 (one nitrogen atom and three hydrogen atoms). Creating ammonia is an energy intensive process, consuming 2% of all the world's energy and generating 1% of the world's carbon emissions. Excess energy can be used to both generate the raw hydrogen as well as provide the heat and pressure required to transform it into ammonia.[33]

EV charging. The transition to electric cars I noted above will create an enormous market for EV charging. Charging companies will seize any opportunity to purchase lower cost electricity, even if it is only seasonally cheap.

A NEW GRID FOR AFRICA

Up to this point in the book, I have focused on using renewable local energy to upgrade the fossil fuel-powered Big Grid. But there is another market where local energy is not just a better option, it is the *only* option. It is hard to believe, but 770 million people have no electricity at all. The vast majority, 580 million, live in Africa.

These are some of the world's poorest families, and many of the problems they face are a result of their limited energy options. Because they light their homes and cook their food with the flames of kerosene, wood, or animal dung, fumes and accidental fires lead to thousands of deaths a year. Education is stifled as children struggle to study after dark. Fresh water is scarce because pumping water with diesel generators is too expensive. People walk for miles to charge a mobile phone. In these communities there are no refrigerators to store food or medicine. Energy poverty is a particularly heavy burden for women, in part because they are most often in charge of the time-consuming task of gathering fuels and are disproportionately exposed to indoor pollution from cooking and lighting.[34]

Country	% of population with electricity	Millions of people without electricity
South Sudan	1%	11
Chad	9%	15
Democratic Republic of the Congo	9%	79
Burundi	11%	10
Liberia	12%	4
Malawi	13%	16
Niger	14%	20
Somalia	18%	13
Burkina Faso	22%	16
Sierra Leone	26%	6
Guinea-Bissau	28%	1
Uganda	29%	32
Mauritania	32%	3
Benin	33%	8
Mozambique	35%	20
Lesotho	36%	1
Zambia	37%	11
Madagascar	39%	17
Tanzania	40%	35

FIGURE 6.2 *African countries where 40% or less of the population has electricity. Source: International Energy Administration 2020 World Energy Outlook (freeingenergy.com/m188).*

Building and operating an electric grid in densely populated areas is very profitable, since there are fewer wires required to reach customers. But in lightly populated rural areas, like much of Africa, far larger investments are required per customer to build a grid. The rural poor in Africa today are in a similar situation to the rural poor in America in 1935, when only 11% of rural farmers had electricity.[35] At the time, the US had the wealth and the vision to pass President Franklin Delano Roosevelt's Rural Electrification Act of 1936, which rapidly expanded the grid's reach. By 1949, 78% of farms had electricity and almost all were connected in the following decade.

It will be more economically difficult for today's African leaders to do something similar. Even when these leaders have the political determination to extend their grids into rural communities, many countries lack the economic wherewithal. Less than 20% of the people in eight countries, including Somalia, Liberia, and Chad, have access to electricity of any kind (see the full list in **Figure 6.2**). Only 1% of South Sudan's population has access to electricity, leaving more than 10 million people with no electricity at all. Across Africa, this has trapped hundreds of millions of people in poverty.

Many African nations are stuck in a dependency cycle with fossil fuels that will be difficult to break. Consider Nigeria, one of Africa's most populous countries. About 40% of the population still has no access to the country's electric grid. With only 4 gigawatts of power available on the grid, the country has become dependent on diesel generators, which together generate 12 gigawatts of power.[36] This cycle is perpetuated by hard-to-change government subsidies of fuel that total almost US$4 billion annually.[37] The politics and the leadership to change this are daunting, but the opportunity to shift some of the subsidies to investments in clean, local energy will be transformative.

With the plunging costs of solar and batteries making local energy viable for these rural families, governments have growing incentives for supporting off-grid electrification. Just a few years ago, off-grid lighting meant little more than small *solar lanterns*— handheld gadgets that combined a miniature solar panel, a battery and a lightbulb. As economies of volume drove down the price of these (first order) components across the globe, innovations on the integrated (second order) systems took off. One light is now three. Radios, flashlights, and televisions are part of many new systems. Innovation across all five orders is taking place at blazing speed.

Realizing the promise of local energy in Africa

Jacqueline Novogratz and Amar Inamdar are two of the most inspiring people I have met on my journey into local energy. When they were planning a trip across East Africa to talk to customers and government leaders about the promise of these small solar+battery systems, they invited me to join them.

FIGURE 6.3 *Jacqueline Novogratz (right) with a Kenyan mother and daughter at their farm. Photo courtesy of Shaiza Rizavi.*

Jacqueline is the CEO of Acumen, a firm she started in 2001 to forge a new approach to solving poverty. She is endlessly positive, tireless, curious, and deeply empathic. She and her team have grown Acumen into a global exemplar of what she calls, "taking a pro-poor patient approach to investing," which is often referred to broadly as *impact investing*. Acumen's manifesto describes this new type of investment beautifully: "It starts by standing with the poor, listening to voices unheard, and recognizing potential where others see despair. It demands investing as a means, not an end, daring to go where markets have failed and aid has fallen short. It makes capital work for us, not control us."

The firm has been a pioneer in making energy-based solutions like lighting and charging affordable in some of the world's poorest regions. Acumen, its people, and the organizations it has invested in have collectively touched the lives of hundreds of millions of people. Jacqueline once shared the words of an Indian woman she knew: "If you have power, there is no wire. Then you are free."

Amar Inamdar is the Managing Director KawiSafi—Swahili for "clean energy"—a Kenya-based venture capital fund created from Acumen's early successes in clean, local energy. Amar and his team are investing in a carefully targeted set of early-stage companies with the strategy of creating a vibrant business ecosystem for off-grid companies that "democratize energy in Africa," as he describes in his wonderful 2017 TED talk.

He has the frame of a runner and speaks with a flowing eloquence that lends depth and gravitas as he speaks. Amar knows the culture, the language, and the people of East Africa because he grew up there. He later earned his PhD from the University of Cambridge while living on the Kenyan plains alongside herds of elephants, studying savanna ecology. After growing his career in global finance, the opportunity to build one of the first clean-energy venture capital firms in Africa called him back home.

He explained the need and the opportunity for off-grid systems:

The national grids are an incredibly expensive and inefficient way to get power to people, especially when domestic energy consumption is low—as it is across so much of Africa. Gross domestic product and electrification are closely correlated. Unfortunately, some folks have a mindset that the only way to get electrification is the grid. We asked ourselves, what if we could get electrification to happen without the expense, infrastructure, and inherent inefficiencies of the traditional grid? This is why we created KawiSafi. There is

FIGURE 6.4 *Amar Inamdar showing a Kenyan farmgirl a photo he had taken of her on his phone. Photo courtesy of the author, Bill Nussey.*

a fast-growing group of companies that offer customers scalable power solutions—energy that is right-sized for their demands and can grow with them to meet their power needs in the future. So, we link electrification and development, rather than the grid and development. And we can help realize this vision by funding amazing young companies and their breakthrough products.

Is it right to use capitalism to help alleviate poverty?

In 2014, UN Secretary-General Ban Ki-moon declared: "Sustainable energy is the golden thread that connects economic growth, increased social equity and an environment that allows the world to thrive. Low-carbon growth can foster decent jobs, empower women, promote equality, provide access to sustainable energy, make cities more sustainable and enhance the health of both people and the planet."

My less eloquent and more business-minded view is that we are watching the creation of a brand-new type of grid that benefits from 21st century technologies. It is being crowdsourced for a tiny fraction of the cost of any other grid in history. It is the quintessential *consumerization* of energy.

Early on, I struggled with the idea of private companies profiting in this market. I thought that every penny available should go toward making these systems available and not be funneled back to investors. But Jacqueline has taught me two important things. First, money from private companies, philanthropies, and government aid is not mutually exclusive. The low-income world can and should benefit from all of it. Second, building the democratized grid with the help of private, high-growth, profitable businesses actually draws in vastly more capital than governments and philanthropists can do by themselves.

Before Acumen launched KawiSafi, it had been successfully using philanthropic-backed patient investment for a decade to build a solar market for low-income people. Acumen was an early funder of several of the most successful companies in the segment including d.light and BBOXX. Acumen's patient capital approach allowed companies to understand the needs of low-income people and build the products and distribution processes that truly suited their needs.

KawiSafi was established in 2018 to scale up Acumen's early successes and build a new energy ecosystem that would solve energy poverty. As Jacqueline explained, "We used private capital to help solve a big problem by going where government and markets had fallen short. Through a combination of philanthropy and private capital, backed by a decade of on-the-ground experience, we helped a new private sector to flourish that was good for people and the planet."

Amar explained the unique role Acumen and KawiSafi play in helping private companies augment and accelerate roles often taken by governments alone:

> The private sector has already proven it can play a vital role in Africa's development. Just look at the cell phone revolution. We've seen mobile money payment systems like M-Pesa transform banking by moving it outside bricks and into the mobile phones of people across the region. Early on, people weren't confident banking could be done this way, but the market proved it could. Experiences like this give us confidence that private organizations and grassroots customers can work with businesses and the government to deliver better solutions than any of them can offer on their own.

Local energy is lifting lives in Africa

BBOXX is one of KawiSafi's investments and one of the largest off-grid providers in Africa. Crunchbase reports that BBOXX has raised $165 million, an impressive amount, particularly for a company pioneering markets that few investors have bet on before.[38] The company describes itself as a next-generation electric utility. To me, BBOXX is a third-order business selling lighting and other electric services. Its success is due, in part, to a proprietary, cloud-connected *solar home system* that combines a solar panel with a battery and control unit. This powers lights, phone chargers, radios, flashlights, televisions—and eventually even small refrigerators. Even more than its technology, I was impressed by the number of local jobs the company was creating. Its Rwandan offices had a well-staffed call center, and a nearby factory employed dozens of people assembling and testing the units. Further out of the city, attendants at a BBOXX retail store signed up walk-in customers. There were even a few BBOXX-branded motorcycles out front that were used by another set of employees to deliver and maintain the units in the remote communities they serve.

Describing these systems in terms of business models and functionality is like describing music in terms of frequencies and sound waves—it entirely misses the impact these systems have on the human experience. So we left the cities and ventured into remote, rural areas—places where the roads are more suitable for walking than driving. We visited people in their homes, and they honored us by sharing their stories.

Sabine is a single mother raising four children—two of her own and two from a relative who passed away. She paid as much as 70 cents a day for kerosene that provided the light for her children's studies and her tiny store. In late 2015, she purchased a solar home system from Africa's largest solar home system provider, d.light (also a KawiSafi investment). It took 136 years for Edison's idea of electric lighting to find its way into Sabine's home. Through the translator, I asked her how it had changed her life. She began talking so quickly, with so much animation and emotion, that the translator gave up. When she was done, he simply smiled and reported that she had used the word "joy" seven times.

The company d.light is pioneer of a game-changing third-order business model called *pay as you go* (PAYG). Sabine paid $25 up front for this $250 system and then 40 cents each day she wanted to use it. Much of East Africa's banking is done through mobile phones; Sabine simply transfers her daily payment and a d.light computer responds with a numeric code. She types that code into the unit's keypad and the system operates for 24 hours. After a few years of daily payments, she will own the unit outright. From then on, her light and charging are free.

Sarah's favorite part of having a solar home system is the rechargeable radio that came with it. I was surprised to hear this and asked her why. She said her village was far from the city and so spread out that she did not know many of her neighbors. The radio was her connection to the world. It was how she kept up with elections, weather reports, sports scores, and world events.

Francis seemed like a man whose life had been even harder than the hard place he lived. With tattered pants and shirt, he began his story by explaining that he had not always been like this. He looked at the ground as he hinted at mistakes earlier in his life that had caused his family to leave him. He said he was getting his life back together and that the solar home system was helping. He pointed to the soot on his ceiling from years of kerosene lamplight. When his children lived with him, the fumes caused breathing issues for them. He hoped electric light would make a healthier and more welcoming home if they came back. Francis told us that he used to walk for hours to get his phone charged, and now he charges it at home.

FIGURE 6.5 *Francis, a Kenyan smallholder farmer, shows off the kerosene lantern that has been replaced by his solar lighting system. Photo courtesy of Shaiza Rizavi.*

But life was still hard. Most days, he explained, he could afford the daily payments, but some days he had to go without. Amar asked him what he wanted after he paid off the system. His answer was definitive. Francis wanted a d.light television that could be run from his solar home system. When he wanted to watch a game, he had to invite himself to a neighbor's house. With a TV, he told us smiling with hopeful pride, he could invite his neighbors to *his* house. What was his favorite part of having the system? Francis said there was a lady living nearby who sometimes visited him so she could charge her phone. He hoped she might take a fancy to him.

The vision to power past conventional assumptions

Even with the best intentions, change comes slowly. All the world's large economies were built with Big Grids. This approach to electrification requires big finance, big equipment, big contracts, and big corporations. It is all too easy to assume this is the only way to electrify a nation. A certain degree of courage is needed to set aside a century of electric convention and think differently. This was exactly the kind of new thinking that was taking place one beautiful evening in a hotel dining room in Rwanda's capital city of Kigali.

Jacqueline had convened a dinner that included some of the most senior people from Rwanda's energy ministry and several advocates for the country's rural communities. Rwanda was and continues to be one of the most forward-thinking countries in Africa. It has been accelerating its development on multiple fronts, and we were gathered that evening to discuss one of the most foundational: electrification.

The Rwandan government had already distinguished itself among African nations by explicitly including local energy—small solar+battery systems, microgrids, and mini-grids—as a core part of its plan to bring electricity to all its citizens. Rather than stick with the conventions of a purely Big Grid architecture and wait another decade or two to electrify the country, Rwanda's economic plan called for increasing local energy from serving 1.5% of its population in 2016 to 22% within a few years.[39] Jacqueline had brought this group together to see if that number should be even higher.

At one point during the evening's conversation, a senior leader from the energy ministry asked us to repeat the costs for these small systems. Jacqueline said that a full system including a TV was about $250 in US dollars. Small lighting-only systems cost as little as $10. The ministry leader responded, "Wait, these numbers are much lower than what we heard when setting our off-grid targets a few years ago. Are we missing something?" We explained that these systems were benefitting from the same type of technology price curves that had made mobile phones so affordable in recent years.

The ministry leader was exasperated. "How can we make long-term policy when the underlying costs are changing so quickly?" he said. It was in that moment that one of the fundamental challenges facing local energy became clear: *Local energy technology moves far faster than policy.* While many industries were accustomed to fast-paced technology changes, this was an entirely new dynamic for the decision makers in Rwanda's energy ministry, just as it is for the large governments across the world, including the US. The ministry leader took all this in and said the opportunity for off-grid systems was larger than they realized and committed to revisit their off-grid targets.

The Rwandan leaders we met that night proved that, given the right leadership, it is possible for policy development to reflect the opportunities created by fast-moving technologies. A year or so after that dinner, I could not have been happier to read that Rwanda had revised its target for off-grid to serve 48% of its population.[40]

Innovating across the value chain

GOGLA, the global trade association for off-grid energy, reports that more than 100 million people have some kind of off-grid system, saving a collective $11 billion in energy expenditures over alternatives like kerosene.[41] Among those, 2.7 million people are using the systems to support an enterprise like charging phones for a fee or operating a shop or restaurant after dark. GOGLA estimates that $5.7 billion in extra income has been generated for these families as a result. BBOXX, d.light, and other companies are elevating their business models into fourth-order platforms and selling specialized appliances like high-efficiency televisions, razors, irons, and even refrigerators.

Consumer conveniences are just the start. Amar explained that one of the next life-changing trends in Africa's economic development is electrification of industry, or what is called *productive use*. He says, "Once people have power, they start using more of it; and once people have productive power—like irrigation, cooling, and manufacturing—their incomes go up, they purchase more goods, and a virtuous cycle of economic development is launched. This feeds into the market, so the demand picture today is a small fraction of the pent-up, potential market that is there to be unleashed."

A similar pattern drove electrification in the US. In the early 1900s, General Electric sold electric appliances like washing machines, toasters, and fans, promising freedom and convenience to millions of Americans. As these devices were purchased, the demand for electricity grew. This drove the need for more power plants. This, in turn, caused the cost of electricity to decline. It was a powerful cycle that electrified all of America by mid-century.

As the market grows and technology costs continue to decline, newer and larger systems will become available. Microgrids will power larger equipment like welders and juicers. Mini-grids will connect multiple homes and commercial operations all at once. These smaller systems are allowing the people of Africa to begin climbing the *energy ladder* years sooner because they are no longer waiting for the Big Grid. The virtuous cycle of increased demand driving down the cost of electrification is already well underway.

||| ||| |||

These last two chapters have given you a small glimpse into the technology, the business models, and, most important, the innovators that are reinventing the electricity industry for the 21st century.

There are far too many innovations to cover in a single book, let alone a few chapters, so I chose to focus on those that will help drive local energy. But even this is only a subset. I am actively tracking more than 50 fascinating areas of innovation that include inverters, digital transformers, fuel cells, carbon capture, artificial intelligence, operations and maintenance (O&M), data and analytics, new financing models, energy efficiency, marketing platforms, modular microgrids, and more. I will touch on more of these in the final chapter.

Collectively, all these innovations, and the hundreds yet to be created, promise to accelerate the transition to clean energy, while lowering the cost to customers and improving the reliability and safety of electricity. And, like Niagara's electricity laying the groundwork for the Wright brothers' affordable aluminum engine, all the follow-on innovations are impossible to predict or even imagine. But all this promise is up against a daunting and tireless nemesis—the monopoly laws that are locking out competition, stalling high-growth capital, and most important, sidelining innovators and entrepreneurs who aspire to build amazing businesses that can change the world.

Many of the electric monopolies want to innovate, and many even believe their handful of high-profile pilots are steps in the right direction. But, for too many of them, such projects are just innovation theater. Nathan Myhrvold put it well: "The problem is like complaining to a railroad company that you can't go 100 mph—in the 19th century, they were going to 30 mph; the people that got to 100 mph built airplanes, and airplanes certainly didn't come from railroad R&D shops."

Just as finding a way to travel faster than 30 mph was inevitable for transportation, so are clean and local electricity systems the undeniable trend in the power business. It is only a matter of how quickly it happens and whether it is built in partnership with incumbent electric monopolies or it rolls over them.

CHAPTER 7

UTILITIES VS THE FUTURE

On April 14, 2020, a shadowy organization called the New England Ratepayers Association (NERA) filed an urgent request with the Federal Energy Regulatory Commission (FERC), the federal department that oversees US electricity. Unlike many other advocacy organizations, NERA does not like a lot of attention, and its lawyers refused to provide much information on their clients or funding sources. But if there was any question about the organization's motives, the final paragraph of the petition said it all, requesting that "the Commission find unlawful, and therefore reject state net metering laws . . ."

NERA was asking FERC to override state *net metering* laws that require utilities to pay full retail rates for any surplus electricity their customers with rooftop solar send back to the grid. In other words, they wanted to slash the money paid for excess solar to the country's 2 million rooftop solar owners, crippling the market for future installations.

The response was overwhelming: 450 organizations, 57,000 individuals and 37 states submitted comments opposing NERA's petition.[1] On July 16, FERC unanimously dismissed it. But the battle was far from over. The dismissal was largely on procedural grounds and two commissioners indicated a willingness to address the issue of net metering in the future.[2] The fate of local energy is at the core of dozens of such public battles, oftentimes against opponents who are as well funded as they are opaque. And these are only the influence campaigns we can witness. Many other efforts are taking place far from our view that we will never hear about.

GOLIATH ROARS

The utility industry and its trade groups were notably silent as NERA made headlines, but they had everything to gain had FERC acted on the petition. Some even wondered if the utility industry was partially or entirely behind the group. If true, this would be just one of a long list of fake grassroots or "astroturf" organizations. A handful of utilities have even been caught funding these fake grassroots organizations which claim to represent a broad set of electricity customers. They are often little more than facades cloaking monopolies' interests in derailing local energy, clean energy, or other existential threats to their business.

This isn't a new tactic in the energy industry. When Samuel Insull sought to influence public opinion in favor of his utility model, he did more than just travel the country making public speeches. He and other electric company owners created the playbook used today. They ghostwrote articles for politicians, they secretly funded groups claiming to be grassroots organizations, and they became heavy advertisers in newspapers willing to favorably cover utilities.[3]

Fast forward to today. Organizations like Iowa's REAL Coalition and Empowering Ohio's Economy have been called out by local journalists as suspiciously opaque, and those suspicions have been strengthened by the fact that these groups often seem to pop up shortly before major pieces of legislation are considered. But few of these astroturfing advocacy groups rival the audacity of Florida's Consumers for Smart Solar (CSS). The group was formed only a year before it sponsored an amendment to Florida's constitution that appeared to guarantee the right to rooftop solar (which Floridians already had) but would have also constitutionally protected utilities' right to charge extra fees to solar owners. Industry watchdog Energy and Policy Institute, reported that CSS was largely funded by utilities and fossil fuel interests, but that did not stop the group's cosplay as a local energy advocacy group.[4]

The group's true motives were embarrassingly revealed when one of its consultants was caught on tape in a private meeting explaining the proposed amendment was "an incredibly savvy maneuver" that "would completely negate anything they (pro-solar interests) would try to do either legislatively or constitutionally down the road."[5]

The solar industry mounted a counterattack and sued CSS to change the amendment's intentionally misleading language. The case made it all the way to Florida's Supreme Court, which declined to force a clarification. In her dissent, Justice Barbara Pariente wrote: "Let the pro-solar energy consumers beware. Masquerading as a pro-solar energy initiative, this proposed constitutional amendment, supported by some of Florida's major investor-owned electric utility companies, actually seeks to constitutionalize the status quo." The ballot measure received more than $20 million in support but was ultimately narrowly defeated in the 2016 election.[6]

To be clear, most utilities do not engage in this type of manipulation. A few even understand the inevitability of local energy and are embracing it. But too many see local energy as a fundamental threat to their business. They believe they are in a fight for their lives, and they are playing hardball.

Jungle warfare

Dark money astroturfing may be the most appalling tactic used by utilities but there are many others, most of which are out in the open and even more effective. Increasingly, fees, permits, and caps are appearing on the power bills of solar owners across the country. Here are a few examples.

Caps. Many utilities cap the size of the solar array you can install—10 kilowatts is a common limit for residential and 100 kilowatts for commercial. A 2001 Georgia law sets a cap on the statewide levels of local energy to an astonishingly low 0.2% of Georgia Power's peak load.[7] This is a tiny fraction of the state's electricity and will have to be raised as solar adoption increases.

Fees. There are too many types of fees to list them all here, but examples include interconnection fees, standby fees, administrative fees, and capacity fees. One of the most creative and egregious is Alabama Power's "backup" fee, which charges customers $5.00 per month for each kilowatt of solar installed on their roof.[8] At $400 to $500 per year in fees for the average home, it undercuts most of the value of rooftop solar, and is most likely the reason why fewer than 200 of Alabama Power's customers had installed rooftop solar by 2019.

Insurance. Solar panels are often covered by homeowners insurance, but some utilities set stringent requirements to dissuade rooftop solar. Florida, for instance, requires $1 million in insurance for systems larger than 10 kilowatts.[9]

Paying to use your own solar. In one of the craziest utility schemes, Maine's gross metering required customer installation of a second meter to measure total solar generation. This allowed Maine utilities to charge a fee, even for kilowatt hours the customer used themselves. It also required the customer to install the second meter, costing upwards of $1,000.[10] It was not until 2019, when the state elected a new governor, that this was overturned.

Connection delays. As solar gains traction, utilities can get backlogged (or drag their feet) and take weeks, or even months, to physically connect rooftop solar and begin paying for any surplus electricity generated.

Mice in the elephant cage

When I worked at IBM, one of my favorite bits of company lore was a story told by a retired SVP, Jim McGroddy, called "The mice in the elephant cage."[11] Small, fast-moving projects (or startups) were the mice. IBM was the elephant. Even though the mice were tiny, they made the elephant terribly uneasy. The cage had plenty of room for everyone, but whenever mice entered the cage, the elephant would inevitably overreact, jump around, and squish the mice.

This story holds true for many large organizations, and it is particularly true of the electricity business. Local energy (mice) evokes deep apprehension in monopoly electric utilities (elephants). These giant companies have never had to share their cage before. Local energy moves quickly, decentralizes control, and accelerates innovation. It creates truly competitive markets, with thousands and even millions of peer-level participants. All of this is foreign to the utilities and makes them uncomfortable. Yes, there is plenty of room in this metaphorical cage for both utilities and local energy, but losing even a little bit of control of their monopoly causes the utilities to start stomping about. Local energy is not a mortal threat to utilities any more than mice are to elephants. Utilities are going to need a lot of help to learn how to share the cage.

It is not hard to see why utilities battle so fiercely to keep local energy out of their cage: They have one of the best deals in the world of business. Their profits are guaranteed by the government, literally. In most places, their profits are based on their investments in power plants and power lines, not on customer satisfaction or even operational effectiveness. In a large part of the US, utilities have no competition, and their customers have no alternative providers.

A century of monopoly protection has allowed the electric utility industry to become disconnected from normal market forces. For example, utilities have *the smallest innovation budgets of any industry,* investing less than 0.1% of their sales on research and development.[12] For comparison, software, pharmaceuticals, and semiconductors all spend 14% or more of their budgets on R&D. At 0.8%, even the paper industry spends eight times more of its budget on research than does the power industry.

Perhaps utilities use their budgets on other areas, like customer satisfaction? It does not appear so. As Peter Fox-Penner shared in his book *Power After Carbon,* ". . . utilities rank lower in customer satisfaction than most other industries. One 2017 ranking from J.D. Power shows utilities near the bottom of twenty-one consumer sectors (with a score of 720), below airlines (with a score of 760) and car dealers (with a score of 820)."[13]

In the absence of competition, utilities direct a large part of their budgets toward the only group of people whose decisions actually affect utility profits—regulators and politicians. Utility spending on federal lobbying ranks third among all industries. At $2.6 *billion* over the last 20 years, only pharmaceuticals and insurance spend more.[14] And this is just federal lobbying, which organizations must legally disclose. Most laws governing utilities are at the state level. Since few states require lobbying disclosures, it is likely utilities spend much larger sums here. The industry seems to have concluded that lobbying is a cheaper path to utility profits than innovation.

Grid defection and the "death spiral"
In 2014, a paper by Rocky Mountain Institute (RMI) introduced a new term into the industry lexicon, *grid defection.* "Equipped

with a solar-plus-battery system," the paper explained, "customers can take or leave traditional utility service with what amounts to a 'utility in a box.' This represents a fundamentally different challenge for utilities. Whereas other technologies, including solar PV and other distributed resources without storage, net metering, and energy efficiency still require some degree of grid dependence, solar-plus-batteries enable customers to cut the cord to their utility entirely."

But lost revenue from grid defection is the tip of the iceberg for asset-heavy electric utilities. Their large, fixed costs will amplify these revenue declines into even larger profit losses. I was alerted to the magnitude of this impact during my dinner with Jim Rogers. He pointed me to a 2013 paper circulated by the power industry's trade group, the Edison Electric Institute (EEI).[15] It described the damage that competition had wrought on airline and telecommunications industries, both of which had been previously protected by a similar form of regulated monopoly. It warned that distributed generation (local energy) would have a similar impact on the power industry.

> The electric utility sector has not previously experienced a viable disruptive threat to its service offering. . . . However, a combination of technological innovation, public/regulatory policy, and changes in consumer objectives and preferences has resulted in distributed generation and other DER being on a path to becoming a viable alternative to the electric utility model.

The paper, which quietly disappeared from EEI's website a few years later, laid out a terrifying vicious cycle that the industry later coined *the death spiral* (Figure 7.1). It starts with a few customers installing solar because it is cheaper than the grid. As customers depart, the remaining customers must pay more to cover the utility's fixed costs. This causes the retail rate of electricity to rise, which motivates even more people to install solar, and so on. Making matters worse for utilities, the cost of solar+battery is going to keep going down, accelerating the cycle.

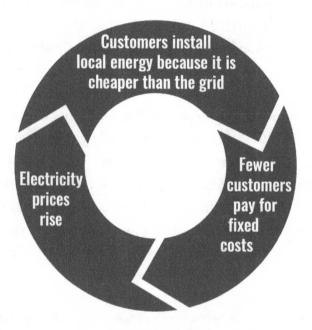

FIGURE 7.1 *A visualization of the electric utilities' death spiral.*

Cost shifting is the monster under the bed

Electric utilities have convinced many regulators that the death spiral represents a mortal threat to the stability of the entire industry. The EEI described it in a 2017 submission to the US Department of Energy saying, "the costs of the [net metering] subsidy are shifted from private solar customers to those customers who do not have, do not want, or in some cases *cannot afford* [to] install private solar generation."[16] [emphasis added.]

This last phrase points to one of the most effective arguments utilities are making to slow local energy and net metering. It invokes the morality of social justice by suggesting that wealthy homeowners installing solar will shift the costs of maintaining the grid to low-income families in the form of higher electricity rates. This *cost shifting* argument is intuitive and powerful, so it is not hard to see why regulators have been swayed.

However, the problem is that the argument is based on a false choice. It implies, somewhat disingenuously, that there are only two groups involved: wealthy households and low-income households. But there is a third group, one that has everything to gain by the debate and one that has more cash than anyone else: the utilities themselves. The power—and the fallacy—of the utilities' cost shifting argument is that it keeps utility profits out of the debate entirely, as if its profits are untouchable. The supposed harm to low-income families triggers so much outrage that local energy advocates forget to look at all the parties involved.

Imagine how consumers would have reacted if Blackberry, the maker of the once ubiquitous smartphone, convinced the government that iPhone buyers had to pay fees to Blackberry to guarantee its profits? Obviously, electric utilities are not regular businesses, and no one wants them to go the way of Blackberry. But as I will explore shortly with a humorous parable, it is absurd that nobody asks whether utility profits should decline when millions of customers want to replace some Big Grid electricity with cheaper, cleaner, and more reliable local energy.

There is another irony to the cost shifting argument: rooftop solar might actually *lower* utilities' costs. How can this be? It has to do with the variable costs of producing electricity. Most people pay a fixed rate per kilowatt hour, but the actual cost of that kilowatt hour to utilities varies across the day. Utilities pay the most to generate electricity on hot summer days when every-one is cranking up their air conditioners. This is called *peak demand* or *peak load*. To meet all this demand, utilities must bring their most expensive peaker plants online. If there were a way to reduce dependence on these peaker plants, utilities could save a lot of money.

As it happens, hot summer days are the very same time when solar panels are producing the most energy. Following this logic, widespread rooftop solar adoption would *lower* a utility's costs by reducing the amount of electricity utilities would otherwise generate from expensive peaker plants. Trimming peak load could also ease the wear on utilities' infrastructure, lowering their maintenance and upgrade costs. One study found that, when environmental and health costs are considered, local solar *subsidizes* non-solar neigh-bors and utilities.[17]

But let us give the utilities the full benefit of the doubt for a moment and analyze the impact on their financials if the cost-shifting argument was 100% true. How badly would utilities be harmed if all residential solar directly undercut their profits? In 2019, residential rooftop solar generated 21 terawatt hours of electricity.[18] That is about one-half of one percent of all the electricity generated that year in the US, or a theoretical $2.7 billion of lost revenue for a $400 billion industry.[19] Is preventing a few billion in revenue declines really worth all the lobbying and astroturfing? It is important to point out that utilities made no efforts to stop a much larger revenue loss the same year. Widespread adoption of more efficient LED bulbs resulted in a $5 billion decline in 2019 utility revenues![20]

The truth is that utilities' costs are incredibly complex and understanding the true economic impact of local energy requires a deep, objective analysis. Fortunately, several independent experts have already weighed in. A widely cited study from the DOE's Lawrence Berkeley National Laboratory (LBNL) concluded the worst-case impact of cost shifting to be about 1% of an electric bill in 2030, or about one-tenth of a cent ($0.001) per kilowatt hour for the average American electricity customer. The study's bottom line was:

> . . . for the vast majority of states and utilities, the effects of distributed solar on retail electricity prices will likely remain negligible for the foreseeable future. [It] will continue to be quite small compared to many other issues.[21]

Could it be this fight is less about *costs* shifting from wealthy families to low-income communities and more about *control* shifting from utilities to consumers and local businesses?

From ratepayers to customers

Realistically, few people will choose to defect entirely from the grid, at least not until batteries are much cheaper. Even then, the grid will remain a convenient form of backup for rainy weeks, not to mention a platform for people to buy and sell electricity between themselves. Of course, if utilities continue piling on fees, they will give customers a good reason to go it alone and grid defection will become a reality.

But even a partial defection by a limited number of people will upend the structure of the power industry, and this is not a bad thing. At a minimum, all of us who pay electric bills will be elevated from our powerless position as *ratepayers* into actual *customers*, the same vaunted and powerful role we have with every other industry from which we purchase goods and services. Empowered with real choice, we will attain the first real energy freedom we have enjoyed since the best option for lighting our homes was candles and kerosene lamps.

DAVID ROARS BACK

The kilowatt hour—the universal unit of electric energy across the world—is trapped. It is locked into a monopoly business model that has barely changed in a century. As things stand today, local energy will not achieve its full promise, and we need to change this. It is time to start a path toward energy freedom, and local energy is an essential first step.

In the following sections, I will explore several important ideas that can free local energy: the social compact, net metering, soft costs, dynamic pricing, new home pre-fit, V2H, local energy marketplaces, PUCs, and a local energy bill of rights.

None of these are a government handout and none require dismantling the electric utilities. They are all fair to all parties. In fact, several of them will directly benefit utilities, too. Most important, I believe all of these are inevitable, so we are wise to begin putting them in place with all haste.

Reasserting the social compact

Utilities serve a vital function, and we are going to need them for years to come. Calls to dismantle them are misplaced. Since the early 1900s, when Samuel Insull popularized the idea of regulated monopolies, electricity has become affordable and universally available. But nearly a century after having achieved his lofty goals, the social compact has eroded. The balance of power has shifted away from customers and toward utilities. In light of new technologies, new environmental risks, and the basic rights of customers, it is time to reassert the social compact. But how can we do it?

Increase transparency. Regulators and state governments should demand complete transparency on costs and other internal utility data. Utilities resist this, citing trade secrets. In an industry with little direct competition, it is difficult for me to see how releasing any of their internal data could result in reduced service levels or increased customer costs. If utilities are a form of public trust, should they not be held to the same level of transparency as the governments that grant their monopoly status?

Curtail lobbying. It is both perverse and deeply ironic to me that profits guaranteed by governments can be used to lobby the very governments that regulate the industry, often to achieve ends that do not benefit citizens of the government. Paid political advocacy and lobbying must be severely curtailed, or at the very least, entirely disclosed, particularly at the subnational level, including details on every recipient. Dark money campaigns have no place in the social compact of monopolies and their customers.

Net metering: getting fair value for your solar kilowatt hour

If defeating local energy is the target for many utilities, then eliminating net metering is the bullseye. Let me explain. Net metering is a policy that pays solar rooftop owners for any surplus electricity they send back to the grid, which is typically 20%–40% of the total solar they generate.[22] In 2021, most states had some form of net metering policy.[23] In most cases, regulators force utilities to pay homeowners the same kilowatt hour retail rate the utilities charge when they sell electricity. It is as if the surplus electricity spins the power meter backward. You can also think of net metering as a virtual battery that is charged during sunny hours and is discharged at night when solar panels go dark.

Rooftop solar owners love net metering because the credits they receive for their surplus solar allow them to pay off their systems many years sooner. Net metering has proven itself to be an essential bridge as solar and batteries become affordable. Advocates of net metering make the same case I did earlier, that reductions in peak load and delays in grid upgrades create a lot of value for utilities. They also point out that more affordable solar spurs more installations, creating jobs and delivering environmental benefits.

Utilities bitterly disagree. They say there are many costs that go into each kilowatt hour they sell—generating the electricity is a small piece of a bigger cost pie that includes distribution, transmission, maintenance, call centers, and billing systems. (Lucrative shareholder dividends, lobbying costs, and corporate profits are also factored into the kilowatt hour price, but utilities do not argue these costs as vocally.)

Regardless of which costs should be factored in, asking what excess rooftop solar is worth to the utilities is the wrong question. It assumes we have a traditional customer-business relationship with electric utilities. But electric monopolies are not like normal companies—they are a government-granted monopoly. Insull's grand bargain creating these monopolies was premised on utilities serving us, their customers, not how utilities can serve themselves.

So the question we should be asking is this: what is your surplus kilowatt hour worth to others in your community? From this perspective, the answer becomes simple. Your surplus kilowatt hour is worth whatever your neighbor would otherwise pay their utility (minus a small fee for transaction processing and using the grid for delivery). The value of our kilowatt hours should never be measured by what it is worth to the electric monopoly but, instead, what it is worth in a fair, open market (energy freedom), if there was one.

Net metering was never meant to be a permanent policy. In future decades, it will be slowly rolled back as people are allowed to sell their excess electricity for what it is actually worth. However, until electricity becomes a fair market, regulators need to honor the spirit of the original social compact and avoid putting utilities' interests ahead of customers' interests. Customers deserve the right to sell their surplus kilowatt hours at market value, either directly to their neighbor or to their utility. To put it another way, if utilities insist they need the monopoly customer lock-in to survive, then it must be two-way, with customers locking in their electric monopolies as buyers at the true value of each kilowatt hour.

We need to recalibrate the way we think about the electricity system and our relationship to the utilities. To be consumers and not just ratepayers, we need to demand more. I will lay out some specific long-term prescriptions for the industry later in the chapter, but first, I want to present several foundational values that should have been in place long before now. These require immediate attention as we begin the inevitable journey toward freeing energy.

Slash soft costs

The costs of building clean energy projects are like a layer cake. At the bottom you have hardware items like solar panels, batteries, electronics, and metal racks. As I have shared in earlier chapters, these costs have plummeted thanks to huge economies of volume. In the middle layer you have installation labor. Even this has been shrinking as new processes and new products reduce the time to install local energy projects. At the top of the cake are soft costs. As you will recall from earlier in the book, these include a laundry list of miscellaneous line items like permitting, utility interconnection, inspections, financing costs, and customer acquisition as well as indirect costs like multiple on-site visits and losing customers who become frustrated with delays. For smaller local energy projects like rooftop solar, RMI reports that soft costs account for 65% of the total installation cost![24] Much of these costs are due to unnecessary bureaucracy and red tape. Worst of all, the US seems to be the only major country suffering from these runaway soft costs.

"Here in the land of technology leadership and free-market enterprise, American regulation has more than doubled the cost of solar," explains Andrew Birch, who has spent much of his career pioneering ways to slash soft costs.[25] Talking with Andrew is a lot of fun. During an interview for my podcast, he was quick to say, "Everyone calls me Birchy." He is sharp, quick, and has the enthusiasm of someone who loves what he does.

Born in Scotland, Birchy received a degree in physics from one of the oldest schools in the United Kingdom, the University of St. Andrews. During his early career in banking and finance, he got a firsthand look at the trajectory of solar costs and knew he needed to get into the industry. So he moved to Australia to get a graduate degree in photovoltaics from the University of New South Wales, the school where Dr. Martin Green's lab famously invents many of the world's cutting-edge solar technologies. Birchy is also a bit of solar royalty. He co-founded and led one of the first generation of large residential solar installers, Sungevity.

Having seen the industry at work in different countries around the globe, Birchy is well positioned to compare policy approaches, and his assessment of the United States is dismal. To understand just how bloated US residential solar costs has become, Birchy compares the market to Australia's:

> Residential solar in Australia has over 20% penetration. So over one in five homes in Australia has a solar system today. If you go solar in Australia, you pay $1.10 to $1.20 per watt. And you get your system within one to three weeks. Compare that to the States, where only one in fifty homes has solar and they are typically paying $3.00 to $4.00 per watt. These are the exact

same panels and electronics. The installations are just as safe. It can take two to three months on average to get that system.

What is going on here? The difference is soft costs. Australia has a single set of national regulations that are reflected in simple, standardized, and automated processes. In comparison, the US is a patchwork of "18,000 jurisdictions and 3,000 utilities with different rules and regulations to go solar," according to the DOE.[26] The Solar Foundation estimates that "Direct and indirect costs of permitting, inspection and interconnection add about $7,000 ($1.00 per watt) to the cost of a typical residential solar energy system." Yikes.

Birchy is helping lead two big efforts to slash US soft costs. The first is a software startup he co-founded, called OpenSolar. The company provides installers with a software application they can use to design, sell, and manage their customers across the entire buying journey. It streamlines proposals and removes the need for site visits, reducing sales and marketing soft costs. OpenSolar has an innovative business model that effectively makes the software available to any installer anywhere in the world for free.

The second initiative Birchy is spearheading is an unprecedented coalition of organizations and governments, all working together to create an online permitting system called the Solar Automated Permit Processing Platform Plus, or *SolarAPP+*. It all started with a workshop Birchy pulled together in early 2018. Since then, the consortium he started has grown to include RMI, NREL, the US Department of Energy, the California Solar and Storage Association, the Solar Energy Industries Association, Underwriters Laboratories, Solar United Neighbors (SUN), and more than 100 local governments who will be using it at launch.

SolarAPP+ is a simple, standardized online platform for installers to register the systems they build and get instant permits. It will automate what used to be a cumbersome physical process of site visits, plan-set drawings, and a protracted back-and-forth with permitting offices. When implemented, this system will standardize the interpretation of codes and standards—making it easy for cities to get on board and support the growth of local energy. In the end, everyone will spend less money and time, from the participating local governments to the installers, to the most important constituents, the homeowners.

"The benefits go a lot deeper than just lowering the price of solar installations," says Birchy. "It also shifts the money you would be paying to a utility into this hyper-productive, economically generative, tax generative, local economy." This includes a lot of new, good-paying jobs. By Birchy's math, the US needs as many as 10 million people installing solar if the country is to achieve its rooftop solar potential.

This stuff is not rocket science. The US managed to pass national regulations that streamlined the installation of rooftop satellite TV dishes.[27] Even the grid itself relies on regulations that unify its technology and control systems. Certainly, we can find a way to unify and streamline the process of installing local energy and slash costs for everyone.

Dynamic pricing: lining up customer prices with utility costs

Most electricity customers in the US pay a fixed rate per kilowatt hour. While appealingly simple, this is not the most economically effective way to charge for electricity. Similar to the economics of airline seats, the cost of generating electricity varies with changes in demand. Dynamic pricing adjusts the price per kilowatt hour over time to better reflect the actual expense of generation. This reduces total costs for both customers and for their utilities. This is not a new idea.

In addition to championing the monopoly business model for electricity, Insull was also famous for promoting dynamic electricity rates. One popular pricing approach is called *time of use* (TOU), where a utility breaks a day into two or three periods, each with different electricity prices. California is rolling out TOU for all residential customers and many other states offer it as an option.[28] Dynamic pricing is a win-win. Customers can shift their consumption, like washing dishes during off-peak hours, to benefit from lower rates and reduce their bills. Utilities' peak loads are reduced and their profits increase.

For example, if your EV takes four hours to charge, a fixed rate of 15 cents per kilowatt hour results in 60 cents per recharge, regardless of when you start it. Given the flexibility, you will probably recharge when you get home from work. Unfortunately for your utility, this is peak load time, when it is paying the most to generate electricity. However, a TOU rate that charges 20 cents from 7:00 a.m. to 11:00 p.m. and 4 cents overnight saves money for you *and your utility*. You simply tell your car to start charging at midnight and the recharge now costs only 20 cents. Meanwhile the utility is *more profitable* because its cost to generate a kilowatt hour is well below 4 cents at night.

Local energy systems take this to an exciting new level. Because solar+battery systems are inherently smart—managing battery charging and grid export as loads and sunlight change—they can easily optimize power flows around time of use rates as well. This intelligent time-shifting can substantially reduce electricity bills.

Pre-build for local energy

When I installed two Powerwalls in my home in 2020, the labor costs were higher than the price of one of the Powerwalls. Yet when I purchased a new clothes dryer, which uses a nearly identical circuit, I had to pay zero labor costs. Why? The largest difference is that most new houses today are pre-wired for dryers, which are an anticipated addition to a home. But solar panels and residential batteries require ripping apart walls and expanding circuit breaker panels. A few hundred dollars invested when my house was being built could have saved many thousands of dollars in labor costs for my local energy retrofit. Builders need to start planning ahead.

I recommend three things: a few extra circuit breakers and associated wire pulled to somewhere near your roof, one or two circuits pulled to a location that could hold a residential battery, like an outdoor utility closet, and a pre-wired outdoor emergency shutoff switch near your power meter so that emergency responders can turn off local energy generation in the event of a fire. If developers added these simple features during home construction, it would significantly reduce the cost of local energy installations.

Some localities are going beyond my modest suggestions, requiring solar in all new construction. Starting in 2020, California began mandating that all new homes include solar rooftops (more recent amendments allow for community solar as well). The state estimates that even with an additional $8,400 in solar costs added, on average, to the price of a new home, homeowners will see their total bills go down $35 per month thanks to lower electricity costs.[29]

Local energy marketplaces

It was April 2016 in the Park Slope neighborhood of Brooklyn, New York, and local energy history was being made. One neighbor, Eric Frumin, sold a few kilowatt hours of his excess solar power to another neighbor, Robert Sauchelli.[30] This groundbreaking moment was one of the few times in a century of electricity that a kilowatt hour was sold without a government-granted monopoly being on one side of the transaction or the other. These two neighbors were the early pioneers in a project called the Brooklyn Microgrid, which grew to encompass more than 50 participating sites before regulatory and cost challenges put it on hold in 2019.[31] This seems like such an obvious thing to do, yet it remains one of the few such programs anywhere in the US.

You can pay me to cut your lawn, or you can buy my used TV. There are flea markets, bake sales, Girl Scout cookies, eBay, Etsy, even lemonade stands. We can buy and sell nearly everything we make or own with anyone. There is just one huge exception—people and businesses cannot sell kilowatt hours to each other due to the grand bargain Insull struck a century ago. Since then, it has been illegal for *anyone* other than a government approved entity, like a utility, to sell electricity. Period. This is the case across all of the US and much of the world. "No other entity is allowed to exchange energy for money," explained Adrienne Smith, the Brooklyn Microgrid's Executive Director at the time, in an article in *Microgrid Knowledge*.[32]

These laws locking down electricity sales were created decades before local energy was even an idea, let alone a cheaper, more reliable alternative. Despite local energy technology improving dramatically each year, these laws governing the grid remain rigidly stuck in time. Of course, utilities are delighted by this.

For innovation to flourish, there needs to be more than a single provider, and more importantly, those providers need the flexibility to innovate with different products and pricing models.

For most industries, competitive markets lead to innovation and better products and services for customers. Unfortunately for the electricity industry, early attempts at embracing competition left huge scars. In 1996, California embarked on an ambitious effort to open up competition for electricity producers in the state. It created a competitive marketplace for power plant owners and utilities to buy and sell wholesale electricity. But on June 14, 2000, the carefully crafted rules for the marketplace that sought to balance the flow of money with the flow of electrons failed spectacularly, and California's grid melted down. The state began months of what would be the largest rolling blackouts since World War II.[33]

As Daniel Yergin explains in his book *The Quest,* to stop the free fall: ". . . the state transferred the financial crisis of the utilities to its own books, transforming California's projected budget surplus of $8 billion into a multibillion-dollar state deficit." This is the same crisis that brought Enron into the national consciousness. The company gamed the newly competitive market to the point of illegality and is blamed by many for wrecking the grid's delicate economic balance and nearly bankrupting the state of California. But Yergin quotes regulatory guru Paul Joskow, explaining that the challenges were due more to ". . . the most complicated set of wholesale market institutions ever created on earth . . ."

Many experts point to all this as a failure of competition, but the truth is that it was just another failure of "big." It is like trying to make someone more athletic by replacing their heart and lungs rather than starting out with better diet and exercise.

Competition is not a panacea; but when properly guided, it forces businesses to balance profits with customer interests and environmental impact. When multiple businesses compete for customers, they necessarily take risks to innovate their products and services. They try to outdo each other in offering the best services at the best price. Not every business survives, but the overall market—and the customers it serves—will thrive.

To avoid more billion-dollar bankruptcies and rolling blackouts, competition needs to be introduced more thoughtfully. Rather than force it into the beating-heart center of a statewide electric grid, competition should be introduced from the outside in—also known as the *grid edge.* This means creating "walled gardens" around small local energy markets. For example, the people living in a neighborhood, or any small community, could buy and sell electricity among themselves. As I will share in the final chapter, *energy communities* like this are already a reality in Europe. This approach allows for hundreds of small experiments, and it isolates the occasional failures.

Net metering is an important start, because it allows local energy owners to sell their surplus electricity. But it is also constraining. There is always only one buyer (the electric monopoly) and the price and pricing model are inflexible. The idea of letting any solar+battery owner sell their electricity to anyone in their community is far more exciting.

The industry has many names for this, including *peer-to-peer trading, P2P electricity, transactive energy*, or my favorite, the *energy internet*, all of which are explored in the final chapter. P2P allows homeowners and local businesses to buy and sell power to each other. This applies to any group of electricity consumers—neighborhoods, office parks, strip malls, small industrial parks, and even towns. The interconnected systems pool resources, increasing reliability and effectively lowering investments and costs for all participants—the foundational elements for a revised 21st century social compact.

Rather than each home or building having its own power meter that measures electricity from the utility, there would be a single meter for the entire local energy community, or what I like to call "pushing the meter up the feeder" (the distribution feeder). Any buying and selling among homeowners behind this meter would be invisible to utilities and exempt from the monopoly regulations. The community could pay its monopoly electric utility to maintain the power lines between buildings and be able to use the grid as a backup, or even as a trading platform to buy and sell with other local energy marketplaces. There are dozens of pilots going on around the world, including a few in the US like the Brooklyn Microgrid and the Vermont Green app being piloted by Vermont's progressive utility, Green Mountain Power. The regulatory patchwork of state electricity regulations probably allows other, similar pilots, but the bureaucracy is dauntingly dense for entrepreneurs and innovators to navigate.

As I will explain in Chapter 10, Australia, Japan, and the European Union have already enacted laws making these local energy marketplaces a reality. It is high time for the US to create a simple national template for local energy marketplaces. It can initially be limited to small communities of a few megawatts—no more than a small neighborhood or factory would use. Because the local energy system would be smart, the regulators can even design how and when grid electricity can be tapped to limit any grid upgrades utilities may claim are necessary to support these communities.

For power industry veterans, P2P electricity sounds like the electricity version of *Alice in Wonderland*—a surreal fairy tale. We have become so accustomed to monopoly electricity that we have all lost sight of just how bizarre that world really is. Like the proverbial frog in the boiling water, years of complacency have led consumers to accept a situation they would otherwise never tolerate.

To give you a sense of how absurd the system has become, it is helpful to imagine if another large industry in the United States followed the same business model as electricity.

A MONOPOLY PARABLE

Our story begins with Ronald McDunald, the colorfully clad and perpetually smiling CEO of a mythical restaurant chain called McDunald's. Ronald was so inspired by Sam Insull's wildly successful regulated electric monopolies that he decided to pioneer the same idea for food. He knew that economies of scale were key, so he abandoned his original idea of building small, local hamburger shops and convinced investors to fund a handful of enormous, centralized hamburger factories. He then built a truck network that could deliver affordable burgers into every home in the US. Aware that regulators were concerned that McDunald's might abuse its position, Ronald submitted to the same government oversight as Insull—a regulated monopoly. McDunald's promised to make hamburgers affordable and universally available. In exchange, McDunald's would be guaranteed perpetual high profits and it would be illegal for anyone else to sell food. In a few short years, hamburgers were everywhere, and no one went to bed hungry. Without any competition, McDunald's became one the most profitable companies in the country, delighting Wall Street investors. What could possibly go wrong?

One day, a scientist came to Ronald with alarming news. "Mr. McDunald, it has recently come to our attention that hamburgers are very unhealthy. We've done some research and discovered there is an alternative source of food called a *salad*. It is much healthier, more nutritious, and believe it or not, it is also cheaper."

Despite Ronald's perpetual smile, he looked skeptical. "Hmmm. I hear this salad thing makes heavy use of this lettuce stuff which, apparently, can be grown only during some parts of the year. If we switched over to salads, people would starve half the year. Give me a decade to run some pilots and I will get back to you."

Ronald thanked the scientist and sent him on his way. Minutes later, Ronald called an emergency meeting with the McDunaldland staff. "Ladies and gentlemen, we are facing the greatest threat in our history. If this salad thing takes off, people will buy fewer burgers and our revenues will decline and our shareholders will be *very* disappointed. People may even realize that they can survive without our hamburgers."

He barked out orders to his staff: "You, go donate money to politicians whose campaigns will defend our hamburger monopoly. You, find some scientists who will publicly declare that hamburgers are perfectly healthy. And you, create a grassroots organization called *Americans for Hamburgers* and pay them to run ads about the high price of salads and the risk of mass starvation."

Meanwhile, people were getting sicker. Public concern mounted. Regulators and legislators finally got involved. Under pressure, but always creative, Ronald turned this challenge to his advantage. Standing next to dozens of politicians, he held a press conference on the statehouse steps. "After a lot of study, we have a breakthrough. We are going to add a piece of this lettuce stuff to each hamburger we make!" Ronald's supporters heralded him as a visionary. Health experts rolled their eyes.

A few years later, the scientist returned to Ronald's office. "Mr. McDunald, I have a new idea that I call a *garden*. People can grow their own salads in their own backyards. It will be much cheaper and much healthier. Think of all the food innovations as people invent endless new recipes in their own homes. What is more American than self-reliance and innovation?"

Ronald realized the depth of a mistake he made so many years ago. The monopoly he negotiated guaranteed him only the exclusive right to *sell* food. It never occurred to him that people could grow their own food for less money than it cost to buy his hamburgers.

Almost overnight, millions of people were growing their own gardens and making their own salads. Homegrown salads took off. Hamburger sales stagnated.

The McDunaldland team poured out across the country explaining to politicians who listened eagerly (hoping for more campaign contributions) that self-grown salads would devastate his carefully managed truck network, bankrupt his factories, and cause widespread increases in hamburger prices. Legislators and regulators were filled with fear, so they were happy to sponsor the laws that McDunald had drafted for them. After all, they did not want their constituents to starve. And besides, McDunald's was the largest political donor in the country—who would want to rock that boat?

Suddenly, garden owners were paying fees for each salad they made themselves. They were charged fees for the size of their gardens. They were not allowed to sell or even give away their salads to their neighbors. Gardening now required permits, which often took months.

Time went by and the scientist had one more idea for Ronald. "Gardening and salads just keep getting cheaper. Gardeners want to form their own little businesses and sell salads to their neighbors. Better yet, we could create this new thing called a *farmers market* where gardeners can sell salads to lots of people at once." Ronald realized these latest ideas would change everything about his business. He told the scientist he had heard enough.

The scientist was crestfallen. How could something as obvious as selling healthy homegrown salads be illegal? Ronald sensed his unease and offered an alternative. "Listen, we care about these health-focused people, so we have decided to start selling salads ourselves. Our massive factories are now cranking out millions of salads. We can make them much cheaper than all those small home gardeners and our delivery trucks save people the trouble of planting gardens."

The scientist asked, "But, if your factory salads are cheaper, why are you selling them at the same price as hamburgers?" Ronald sighed. This scientist clearly did not understand business, so he spelled it out: "Well, we have to invest a lot to make these new salad factories, and of course, our truck network is very expensive. And besides, who really wants to grow their own salads when they can just buy them in a nice foam box with free plastic forks?"

And there you have it, dear reader. Replace hamburgers with fossil fuels, salads with solar power, and gardens with rooftop solar and batteries, and you pretty much describe the electricity business today. So, how do we even begin to reinvent this industry?

Meet the 201 people that control 72% of the US electricity industry

Most people are surprised to learn that the electricity policy in their state—actually, every state, as well as Puerto Rico and Washington, DC—is controlled by a tiny group of people they did not know existed. These groups go by names like *Public Utility Commissions* (PUC) or *Public Service Commissions* (PSC). PUCs directly regulate the large *investor-owned utilities* (IOUs), which account for 72% of US electricity and the more than $300 billion we pay them each year.[34] They set nearly every policy that affects your electric bill and your ability to use local energy. According to Ballotpedia, these commissioners are elected in 11 states and appointed, usually by a governor, in another 39.[35]

Just as the Federal Energy Regulatory Commission (FERC) takes its direction from the federal legislators, the PUCs take their direction from the legislators in each of their states. Most PUCs also oversee telecom, natural gas, and water, but electricity policy is the most contentious part of their jobs these days. Most of the commissioners' time is spent with the large investor-owned utilities in their state and, increasingly in recent years, representatives for environmental and consumer advocacy groups. One group they almost never hear from are the very *ratepayers* the commissioners are paid to protect. They need to hear from you.

The commissioners in my home state of Georgia are elected. They appear on the ballot every few years. I am embarrassed to say that until I started writing this book, I knew nothing about their role and just glanced over their names on the ballot.

My friend James Marlow is a true solar veteran and active in both state and national solar policy. He offered to take me downtown and introduce me to the commissioners. "You can do that? You can just go and see them?" I asked. James laughed. "Yes, of course." Georgia is home to one of the country's most well-known commissioners, Lauren McDonald Jr. Known to everyone as "Bubba," he is a big man with a powerful handshake and enough charisma to fill a stadium. He is widely credited with bringing the solar industry to Georgia— or as I like to say, turning our red state green. This is all the more surprising since Bubba is a die-hard Republican. Die-hard.

I asked Bubba why he got behind solar when most other Republicans in the South were shunning it. He answered, "I knew a secret that no one else knew." He paused. I was transfixed. "Yes?" I asked.

He leaned in, lowered his voice to a whisper and said, "I knew that solar was cheaper." While Bubba was being theatrical and humorous, at that moment I realized just how important it was that all of us reach out to our commissioners. While Bubba's point about solar being cheaper was common knowledge among clean energy advocates, few other leaders in his position see it as clearly as he does. Many remain hamstrung by misinformation, and some are still actively skeptical.

Who do you talk to if your electric utility is not an IOU? For those of you served by an Electric Membership Cooperative (EMC), you are in luck. You may be surprised to know *you* are a part owner and full-on voter for your EMC. You can even run for the board of directors if you like. If you are served by a *Municipal Electric Utility* (muni), then welcome to one of the fastest growing segments in the utility industry. Your electricity is controlled by the same elected officials who set policy and provide many other local services in your town, county, or city.

It is up to all of us to make sure our commissioners, our state legislators, our mayors, and our governors know about local energy and the benefits it can bring to their communities. If you are inclined to reach out to any of these officials but are not sure what to press for, feel free to borrow from the Local Energy Bill of Rights I lay out next.

THE LOCAL ENERGY BILL OF RIGHTS

Over one hundred years ago, the US government was wrestling with a crucial question: what is the best way to deliver safe, affordable electricity to everyone?

Given the limited options available at the time, they struck a grand bargain with Insull and other utility leaders. The government exempted the utilities from competition in exchange for a promise to electrify the country.

Fast forward to the 2020s. The world is radically different. Amazing new technologies exist. It is time to revisit century-old assumptions and ask the question again. As voters work with governments to find answers appropriate for the 21st century, here are five foundational principles that need to inform the discussion. Each of these rights supports and enables local energy, offering choice to consumers that will ensure a far more balanced answer for the next century.

1. **Self-reliance.** We have the right to produce, store, and manage our own electricity.

2. **Community empowerment.** Our neighborhoods and local communities have the right to manage our own electric distribution and have a single metered connection to our utility.

3. **Fair value.** We have the right to buy and sell electricity between our neighbors and within our communities. As long as this right is withheld, utilities must pay for the electricity we put back on the grid at the same rate our neighbors would pay us for it.

4. **Balanced interests**. Utilities shall not impose fees or processes that unfairly impede our right to local energy.

5. **Energy equity**. All individuals and groups shall have access to local energy, including fair and equitable financing from private and governmental sources.
   (Thanks to SUN and its Solar Bill of Rights for inspiring this idea.)

REINVENTING ELECTRIC MONOPOLIES

Too big to fail (and too important to live without)

Many local energy enthusiasts fantasize about a world without electric utilities, but that is simply not practical. Far be it for me to dash anyone's dreams of energy independence, but the Big Grid remains essential in many ways. For example, factories and industrial facilities consume 25% of US electricity and large centralized power plants will remain the least expensive source of electricity for them for many years to come.[36] Moreover, some particularly energy dense industries like steelmaking simply do not have the rooftop space to power themselves with solar.

Similarly, many dense urban cities lack the surface area to capture enough solar to power themselves. In his book *Power After Carbon*, Peter Fox-Penner compares the potential for local energy in sunny Phoenix with less-sunny New York City. Phoenix is a city of 1.6 million people with 3,349 people per square mile. If you covered every available rooftop with 16% efficient rooftop solar panels, you could provide only 58% of the electricity the city uses. Even when covering parking lots, roads, and walls with solar, and assuming futuristic 28% efficient panels, Phoenix just barely powers itself.

New York City has a population of 8.4 million people, with 27,000 people per square mile. Eight times denser and much less sunny than Phoenix, New York City can satisfy only 17% of its own power consumption with local solar, even with the most generous technology and surface coverage assumptions. Clearly, when it comes to keeping our cities lit and our factories running, local energy can play a role, but we will still need large (renewable energy) power plants for the foreseeable future. Besides, the edge markets I describe above cannot work without the millions of miles of distribution power lines that are already connecting our homes and buildings today.

Insull argued that electricity is a *natural monopoly,* that economies of scale mean a single, giant company is the best option to provide affordable, reliable electricity. I have argued that this is no longer the case for power generation. But when it comes to building and maintaining distribution infrastructure, I believe his argument remains true. The costs and chaos of competing companies pulling wires to your house or apartment building makes no sense. Whatever else changes, the regulated monopolies are likely to continue managing the millions of miles of wires to our homes and buildings. And that is okay.

But as local energy becomes more affordable, it will become the cheapest and most reliable way to provide most of the power to the homes and buildings of the 70% of Americans who live in suburbs and rural areas.[37] The COVID-19 pandemic and the associated work-from-home movement is spurring millions to leave cities for the greener pastures of suburbia, further increasing the market for local energy.

How can today's electric utilities survive the transition from ubiquitous electricity sellers to power line caretakers? How can we ensure they avoid the fate of the US Postal Service, a once essential institution that is now trapped between rapidly evolving technologies and slow-moving legislative responses that oversee it? In the next section, I will lay out a high-level roadmap, borrowing liberally from experts who have spent careers working on this.

The future of electric monopolies: smart integrators

In *Power After Carbon*, Peter Fox-Penner offers an incredibly thoughtful set of options for utilities as they navigate the unstoppable transition to both renewable energy (solar, wind, and storage) and distributed generation (local energy). Like other monopolies before it—railroads, airlines, natural gas, telecom—utilities must be recast for 21st century technologies and priorities. In my simplified view, utilities should transition from selling electrons to operating electron marketplaces. They should become fourth-order businesses, providing a platform for thousands of companies and millions of prosumers to trade electricity and related services. Peter calls this utility model *Smart Integrators* (SIs), and it seems well suited for a future with widespread local energy.

He explains that becoming SIs allows utilities to build on their strengths and avoid the risk of having to master the new world of marketing and direct competition. As SIs, traditional utilities become the backbone or switchyard of the electricity market rather than the storefront. Many aspects of this business model—for example, managing distribution power lines—will continue to be a *natural monopoly* and benefit from the processes and people utilities have put in place over decades.

The big change in this new model is that day-to-day marketing and customer relationships are handled by a new set of organizations he calls Energy Service Companies (ESCOs)—the energy-equivalents of Google or Amazon, serving customers via "individually curated customer experience[s]."

More than a dozen states in the US have already taken baby steps toward this with *retail choice*. This allows customers to choose from a set of competing providers of retail electricity. SIs could take this much further, to include novel new pricing plans specifically designed for prosumers, or *energy-as-a-service* where customers pay for lighting and heating rather than kilowatt hours.

SIs and ESCOs represent one new option for electric business models, but there are many alternatives and countless variations that will evolve across each state and local regulatory regime. While utility revenues may decline in many of these models, shifting into this fourth-order platform business model opens the door to the higher profits of other marketplaces like stock exchanges and e-commerce platforms.

The future is arriving ahead of schedule

The economic juggernaut of local energy is coming much faster than anyone is prepared for. As ideal as it would be, there is no time to design a utopian smart grid or craft idealized policies, hoping to avoid the chaos and risk of rapid change. As the price gap between grid electricity and local energy expands, debates over net metering will become less necessary. We can think of Hawaii's experience with this type of rapid change as a postcard from our future. In 2013, a surge in solar installations threatened the stability of the island's grid.[38] In response, regulators essentially disallowed new solar. The outcry was deafening. Policymakers were forced to get creative. In Chapter 10, I describe the island's common-sense policies that embraced batteries and redefined the utility's role as a backup provider. Local energy cannot be stopped—it can just get smarter.

Fortunately, even though today's passive, "dumb" grid is not ideal, it is more than sufficient for the policy changes I have outlined. Simpler systems can be surprisingly effective, even if their evolution is not what engineers would initially design. This, in fact, happened with the evolution of computer networking and the internet. In the

1980s, there were many efforts to create a standard for interconnecting computers in a network. Most were designed with the "brains" in the network itself, making the hardware connecting computers expensive and complex. Ultimately, it was the comparatively simpler Ethernet architecture that prevailed. Created in the 1970s by internet pioneer Bob Metcalfe, early Ethernet allowed for much simpler, cheaper networks by putting much of the brains in the computers and servers, requiring little or no network-specific gear. No one, including Bob, could have imagined that Ethernet would evolve to become the backbone of the internet, stitching together tens of billions of devices. Remarkably effective systems can be built without top-down control.

Think about flocks of birds, swarms of fish, or ant colonies. These animals work and move together, often quite beautifully, without any notion of top-down management and control. The lessons from Ethernet and from Mother Nature can teach us a lot about designing the future of the Big Grid. For example, building on pioneering systems like the Stone Edge microgrid I covered earlier in Chapter 3, "The Rise of Local Energy," millions of "smart" solar+battery systems can be programmed to coordinate and play nicely with the Big Grid, improving stability and lowering costs for every part of our electric future.

And if you think the utility industry is so entrenched and politically powerful that it can prevent the adoption of local energy, consider what has happened to the coal industry. In 2016, coal companies enjoyed unprecedented support from then-President Trump, who rolled back dozens of environmental regulations and enacted several supportive policies. This was to be coal's big comeback. But three years later, the industry was in retreat. Demand was shrinking. Investors and banks refused to continue funding the industry. Every large US coal company declared bankruptcy. Yes, coal is dirty and environmentally damaging, but it was coal's high costs that ultimately prompted the shutdown of more coal plants than had been retired during the last four years of President Obama's supposed "war on coal."[39]

No matter how much special interests spend on lobbying, they cannot withstand the pressure from consumers and businesses that will be unwilling to pay two or three times more for electricity they can generate themselves or buy from someone selling it on a local market. The political winds will shift. Bipartisan division will give way to the one universal value shared by politicians of both parties—expanding the wallets of the citizens that elect them. And that is exactly what local energy is doing.

There is maybe a 10-year window where forward-thinking utilities will have a chance to embrace local energy and become a bedrock part of the new energy internet in their regions. Unfortunately, most utilities that fight the change will be relegated to simple, low-margin caretakers, most likely absorbed into their more forward-thinking brethren.

(III) (III) (III)

A war is brewing. It is big versus small. Control versus choice. Powerful corporations and entrenched government bureaucracies versus individuals and communities. The old business model is a dinosaur—slow, cold-blooded, seemingly invincible but vulnerable to rapid change. Local energy is the age of mammals. Smaller. Faster. Self-regulating and adaptable to different conditions.

In nearly all ways, the future of electric monopolies is in their own hands. They can embrace the coming changes and play an essential role in our clean local energy future. Or they can resist change, foist unfair fees on local energy owners, hire more lobbyists, and when all the dust settles, still find themselves on the wrong side of history—sidelined and irrelevant.

I believe the best role for local energy champions like us is to do everything we can to help the electric monopolies find their way to a future where everyone wins.

The battle for local energy is playing out amid a broader and more intensive war, one fought to slow the adoption of solar and batteries. Advocates for the fossil fuel industry, with help from a second, somewhat surprising industry, are amplifying serious concerns about renewable energy in the media and throughout the halls of regulators and legislators. Some smart and influential people would have us believe that solar and batteries are expensive, environmentally damaging, and an enormous, failed science experiment. How much truth is in their arguments?

In the next chapter, I will share what I have learned, starting with a cab ride in Las Vegas.

CHAPTER 8

THE BATTLE FOR PUBLIC OPINION

As I stepped into the taxi, the driver introduced himself as Steve.

"What brings you to Las Vegas?" he asked.

I explained that I was attending my very first solar conference, Solar Power International. With 20,000 attendees, it was the largest such event in North America.

To my surprise, Steve turned out to be an expert on solar. Or at least, that is what he told me. As we drove into town, I became a one-man congregation for Steve's sermon on the perils of renewable energy.

"All those 20,000 people are wasting their time," he explained. "Everyone knows that every solar company is losing money. They only survive because of massive subsidies from the government."

I responded, "Steve, you know I'm in that industry, right?" Steve took my response as an invitation to continue pontificating. Obama's war on coal had wiped out countless American jobs; there was not enough land for all the panels we would need; and mining for the raw materials used in solar and wind would wreck the environment. He went on for most of the ride.

That cab to my hotel was my first immersion into an alternate reality where clean energy was a boogeyman forced upon the country by liberal elites. At first, I assumed Steve was part of a fringe group that also wore tinfoil hats on the weekend. After all, nearly everyone I know was either supportive of clean energy or, at worst, ambivalent toward it. But as we have learned during the COVID-19 pandemic, even simple ideas like mask-wearing can become wildly divisive. As I researched the anti-clean narrative, I discovered Steve's views were anything but fringe. Anti-solarism is a *very* large and thriving community full of conspiracy stories and endless streams of easy-to-remember sound bites.

Even the most outrageous myth has a kernel of truth in it—that is what makes misinformation so difficult to debunk. But evaluating the costs and impacts of different energy sources, while not a pure science, was nevertheless something that could be addressed through a serious, fair-minded analysis. Was there any validity to the arguments that Steve (and not coincidentally the fossil fuel companies and utilities) were making? This set me on a multi-month quest to separate the trickles of truth from a firehouse of misinformation.

Most of the myths fell apart under light scrutiny. But somewhat surprisingly, a few of the persistent anti-clean-energy narratives proved to be truer than I expected. This chapter is about my journey into the battlefront of the story wars between two different centuries of energy—20th century fuel-based power plants versus 21st century clean energy.

The myth makers

For many people, renewable energy is like puppies and rainbows—the more we have, the happier they are. But there are others who view renewable energy as the end of days. For coal and natural gas producers, a transition to solar and wind amounts to a death sentence. US coal mining has already shrunk from a $52 billion industry in 2011 down to a $17 billion business in 2021.[1] About 90% of coal mined every year is used to power our grid.[2] About one-third of the output of the $166 billion natural gas industry is used to generate electricity.[3]

Even nuclear, a rare point of agreement between the 20th century energy incumbents and some climate advocates, is fighting for its future as solar, wind, and batteries continue their stunning cost declines. And it is not just fuel producers who are fighting the future. Electric utilities have thousands of fuel-based power plants that are scheduled to run for decades. If renewable energy keeps growing, these older plants will be retired early, costing the utilities billions of dollars.

As Leah Stokes explains in her book, *Short Circuiting Policy*, "These companies win by stalling—by keeping their coal plants open, or by digging up more fossil fuels. Delaying policy action provides these companies with money." By spinning myths and leveraging a century of political influence, these companies have ". . . dragged legislatures, bureaucracies, the public, the parties, and the courts into a debate over climate science and the need for clean energy. And on balance, it is a debate they have won."

There are too many myths to cover, so I will focus on just the four most widespread: solar is too expensive, solar takes too much land, solar requires too many raw materials, and solar panel waste creates an environmental mess. I will also explore another commonly held conception that is based in some truth: the idea that solar's intermittency makes it expensive, at least for now. Finally, I will wrap up with a quick look at the most contentious and myth-filled topic in energy: nuclear power.

MYTH: SOLAR IS TOO EXPENSIVE

My cab driver Steve began his rant where so many others start—solar costs and subsidies. The argument goes something like this:

Solar costs a fortune. Without government subsidies and tax breaks, solar would have no traction at all. Billions in taxpayer money are being wasted to prop up this uneconomic "science experiment." Multiple studies prove solar's high costs by showing that increased adoption of solar and wind drive up electricity prices.

Let me unpack this.

Every kind of energy is subsidized, not just solar

In 2005, Republican President George W. Bush signed the Energy Policy Act into law. A critical part of this sweeping legislation was the Investment Tax Credit (ITC), the first major tax incentive for solar energy. The ITC works by reducing tax payments by as much as 30% of the cost of a solar project. It was a popular bipartisan proposal at the time, evidenced by the fact that Bush extended the ITC several times while in office, and it was extended again in 2015 under President Obama.

For critics, however, the ITC, along with its sibling for wind, the Production Tax Credit (PTC), has been a political punching bag for years. Opposition to these tax credits is usually based on the idea that they represent "the government picking favorites." It is time, critics say, to eliminate them and let "the market decide."

This almost makes sense, except for one critical reality—the US has been subsidizing every kind of energy for decades, including coal, oil, natural gas, and nuclear. In fact, from 2006, when the ITC went into effect, through 2018, solar projects received $31 billion in tax subsidies, wind projects received $41 billion, and fossil fuels received nearly as much as both of those together, $68 billion.

The goal of tax incentives on energy investments is simple—they help nurture promising energy technologies for a period of time as they scale up. When the incentives ultimately expire, the best technologies will have achieved commercial viability and stand on their own. Though both the ITC and PTC have been extended a few times, they were both designed to expire.

But some tax incentives get extended indefinitely. A tax break created in 1926 called *percentage depletion allowance* was established to encourage coal, oil, and gas projects. It allowed the fossil fuel industry to save more in taxes than the cost of the original properties being mined or drilled. Investopedia says, "the depletion allowance has made oil and gas . . . one of the most tax-advantaged investments available."[4]

According to annual data published by the Joint Commission on Taxation (part of the US Congress) and other sources, fossil fuels have received a whopping $214 billion in federal tax subsidies from 1978 to 2018, compared with just $42 million for wind and $31 billion for solar. A 2017 report from the nonprofit advocacy organization Oil Change International looked at just 2016 and found fossil fuels enjoyed $5.8 billion in state-level incentives per year, $14.5 billion in consumption-side incentives per year (like helping people buy heating oil), and $2.1 billion per year to subsidize overseas fossil fuel projects.[5]

According to environmental advocates, these subsidies are only a fraction of the benefits enjoyed by the fossil fuel industry. They assert that current fuel-based electricity rates are artificially low because many costs to the environment and human health are not accounted for or are being deferred to future generations. For example, lining coal ash ponds, restoring coal strip mines, and decommissioning gas wellheads will easily cost hundreds of billions of dollars. Few of these costs are reflected in our electric bills. Furthermore, advocates argue, indirect costs are even higher. A 2019 paper published in the National Bureau of Economic Research found that indirect costs such as premature deaths, health care, and climate impacts from fossil fuel electricity production were $133 billion in 2017 alone.[6]

What is the cheapest way to generate electricity?

In late 2018, I was presenting to a group of Fortune 500 executives. To kick it off, I asked for a show of hands on whether coal or solar was cheaper. Since you are reading this book, you already know the answer, but this group did not. Nearly all the hands voted that coal was cheaper. That was wrong in 2018 and it is even more wrong today.

Comparing the costs of generating electricity
(2020 leveled cost of electricity in US cents per kilowatt hour)

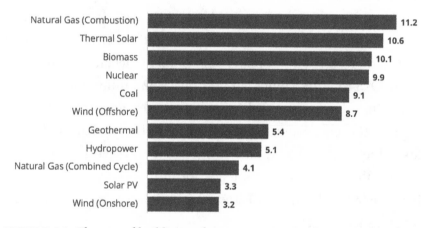

FIGURE 8.1 *The cost of building and operating various electric power plants in 2020. Sources: EIA, NREL, LBNL, Wood Mackenzie, and Lazard (freeing energy.com/g108).*

Compiling the data from six well-respected sources shows that photovoltaic solar (3.3 cents per kilowatt hour) is cheaper than coal (9.1 cents), nuclear (9.9 cents), and even natural gas (4.1 cents). In 2019, only wind was less expensive (3.2 cents). But even wind is predicted to come in second to solar's unstoppable price declines. In fact, solar is so cheap, it costs less money to build and operate a brand-new solar plant than it does to operate an existing, fully paid-for coal plant.

It is not hard to find studies claiming that electricity prices go up in regions that aggressively adopt solar. Germany, a particularly early adopter of rooftop solar, is one of the most common examples cited. When reviewing these studies, make sure to look closely at the data and see how much of the solar was adopted before 2016. Why? Because solar *was more expensive* back then. It is like adding a few mansions to a modest suburban neighborhood and being surprised the average house price goes up.

Now that we are in the 2020s and solar is one of the cheapest ways to generate electricity, it only makes sense that embracing it will drive *down* the costs of operating the Big Grid. It is like adding a few tiny homes to that same suburban neighborhood—of course, the average price of the homes will go down.

As I explained in the previous chapter, the determination of your Big Grid electric rate is incredibly complex. It is a mix of new construction, stranded assets, politics, maintenance costs, and more. Some critics argue that high penetrations (higher than today) of intermittent renewables makes managing the grid more complex, which can add costs. As I will present shortly, advocates argue that modern technology provides multiple ways to address intermittency and any cost impacts it presents are short-lived. In all cases, claiming a direct cause and effect with early solar penetration is a misleading oversimplification.

MYTH: SOLAR TAKES TOO MUCH LAND

Another common argument against solar takes a shot at all the space needed for panels. The argument is misleading, and it goes something like this:

> Solar takes too much land to be practical. Some experts say that powering the US with solar would take 25%–50% of

the country's land. Growing the solar industry means cutting down forests, destroying animal habits, and taking land away from farming and agriculture. And what about dense urban areas? Even if every roof and street were covered with solar, most modern cities are too dense to be powered by solar panels. Instead, we need to focus on nuclear and natural gas, which are hundreds of times more energy dense and require relatively small amounts of land for the same energy output.

Solar can power the entire US with under 1% of the land

Let me start by saying it is absurd to suggest that 25%–50% of US land would be needed to power the country with solar. Where do people come up with this stuff? The actual number is well under 1%. There are several ways to do this math. Rather than use theoretical numbers based on watts per meter squared, let us look at real-world data from existing solar farms and extrapolate those to total US electricity consumption.

The US consumes roughly 4 trillion kilowatt hours of electricity each year (that is 4 petawatt hours or 4,000,000 gigawatt hours for those keeping score at home). The most land-efficient solar installations generate more than 500,000 kilowatt hours per acre per year. This means the US can be powered with 8,000,000 acres, or 12,500 square miles of solar plants. (For the haters, I will readily acknowledge that this is just a fun math exercise—a truly solar-powered country would require a lot more storage and power lines than we currently have).

Thousands of square miles may still seem untenable, but the US has 3.8 million square miles of land, and less than one-half of a percent is needed to power the country with solar. To understand just how achievable 12,500 square miles of solar panels might be, consider other large land uses in the US:

- 40,200 square miles—the land leased by the oil and gas industry.[7]

- 13,125 square miles—the land that has been impacted by surface mining coal.

- 49,300 square miles—the land used to grow corn for ethanol.

- 17,100 square miles—the estimated surface area of US roads (8.8 million lane-miles at an average 10-feet wide).

- 49,400 square miles—the land used for lawns.

10,000 square miles of solar panels can power all the US

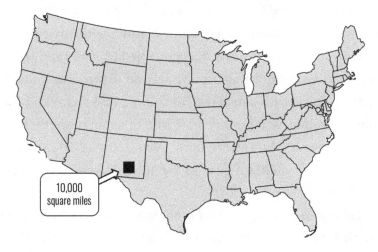

FIGURE 8.2 *An image of 100 miles by 100 miles juxtaposed on a US map. Courtesy: mapchart.com.*

As a thought exercise, consider if we put all the solar needed to power the country within a single block in the sunniest region of the US. This works out to be a 112 by 112-mile square located somewhere in the Southwest US. Most of the examples of land-efficient solar plants are several years old. If we assume our theoretical new country-powering solar plants use modern, higher efficiency panels, that square of land shrinks a little bit and matches the numbers Elon Musk told the National Governors Association in 2017.[8] He said:

> If you wanted to power the entire U.S. with solar panels, it would take a fairly small corner of Nevada or Texas or Utah; you only need about 100 miles by 100 miles of solar panels to power the entire United States. The batteries you need to store the energy, to make sure you have 24/7 power, is 1 mile by 1 mile. One square mile. That's it.

Solar does not need to replace farmland—it can make farming better
Will rapidly expanding solar farms conflict with agriculture? Can the rush to build solar displace valuable farmland and reduce the crop and livestock production we rely on to feed our families? Let me answer this with data. Over 40% of the US is farmland—a stunning 1.4 million square miles.[9] While in some cases there will be conflicts over the best use of land, overall, the 12,500 square miles of solar barely makes a dent in the land dedicated to farming in the US today.

There is so much farmland in the world that even the small fraction that has been abandoned would be enough to power the planet with solar. A 2020 study found that more than 300,000 square miles of global farmland have been abandoned over the last few decades.[10] The authors determined that two-thirds of this land is well suited for solar, more than enough land to produce the 22 trillion kilowatt hours of electricity the world consumed in 2018.[11]

In fact, early research shows that solar farms and food farms may be more complementary than conflicting. A wave of new projects called *agrivoltaics* are demonstrating that solar and farming can not only coexist, they can mutually improve each other's productivity. In one Oregon project, grazing grasses raised under solar panels grew 90% more thanks to reduced evaporation.[12] Another project in Arizona found that cherry tomatoes grown under solar panels doubled their fruit—with no additional water.[13] Even better, the tomatoes lowered the panel temperatures by 9°C, increasing solar power output by several percentage points.

FIGURE 8.3 *Sheep help keep the grass trimmed at a town-sized microgrid in Minster, Ohio. Courtesy of Don Harrod.*

About half the US farmland is pasture, which is used for grazing animals like cows, sheep, and goats. This type of agriculture is particularly synergistic with solar. Not only do solar panels provide welcome shade for the animals, but their grazing also reduces one of the largest costs in operating solar farms—vegetation management.

Dual use puts solar to work without requiring pristine land

Earlier in the book, I discussed one of solar's most overlooked advantages and one of its four superpowers, dual-use. By installing solar on top of existing structures or in places with no alternative use, we can slash the amount of land needed for new solar. Together, residential rooftop, commercial solar, floating solar, and parking lot canopy solar represent an enormous potential footprint—almost enough to power the entire US. Of course, the point is not that all these dual-use applications need to achieve their full potential, only that there are nearly endless locations for solar that do not require covering arable fields, farms, or forests with solar panels.

The best part about dual-use solar is that it embodies the core idea of this book, *local energy*. This means that the electricity is generated near where it is consumed, reducing the reliance on long-distance power lines, which are expensive and failure-prone in severe weather.

I will offer one final thought on land. For those groups advocating that farmland is too precious to be wasted on energy production like solar, I suggest they point their lobbying guns in another direction. There is already a truly massive landgrab underway for another kind of energy. It is the most expensive large-scale way to generate energy and it survives only because the US government mandates it. If we repurposed its land use for solar, we could power the entire US and much of Canada. What is this despoiler of energy markets and future solar silver bullet?

Use sunshine for solar panels, not ethanol

In late 1973, the OPEC oil embargo brought America to its knees, exposing the country's deep addiction to foreign oil. Determined to become energy independent, the US government turned to the country's massive agriculture industry for help. How do you reduce dependence on OPEC oil? You grow it. The ethanol industry was born, and corn was the crop that would feed it. To turbocharge the ethanol industry, from 1978 to 2018, a variety of subsidies for renewable fuels were created, totaling more than $80 billion, which is more than solar and wind combined over that time.[14] While most of those direct subsidies have expired, the ethanol and corn industry still benefit from federal laws that require nearly all US gasoline to contain 10% ethanol.[15]

Corn has become a mega-business. Each year US farmers plant about 140,000 square miles of corn, of which 30%—42,000 square miles worth—is used to produce ethanol![16] (Another 40% is used to make animal feed, and less than 2% is directly consumed by people.[17]) While farmers benefit from this government-created market, corn ethanol is one of the least efficient ways to generate energy. So what if we repurpose our corn fields and use the land for solar farms instead?

Photosynthesis in corn converts less than 2% of sunlight into useable energy versus modern solar panels, which convert 10 times that. Internal combustion engines convert only 20%–30% of ethanol's energy into miles, while the motors in EVs can convert 90% of electric energy into miles, which is three or four times more efficient than gas mobiles. When combined, an acre of solar panels can move an electric vehicle 70 times farther than an acre of corn can move a combustion engine car.[18] It is not even close.

The biggest beneficiaries of replacing corn with solar are the farmers themselves. With family farms going bankrupt at record rates, solar offers a better deal than corn.[19] It is not unheard of for a farmer to make two—or even three—times more money per acre hosting solar panels than planting crops.[20] Solar also produces a reliable stream of guaranteed profits, an ideal alternative to the ever-changing government support programs and volatile crop prices.

If even a fraction of the land used to grow corn for ethanol were upgraded to solar, the US would be well on its way to a truly clean energy future. Additionally, we would reduce fertilizer runoff, dramatically reduce greenhouse gas emissions, and provide American farmers with stable incomes that do not cost taxpayers a dime.

MYTH: NOT ENOUGH RAW MATERIALS

The raw materials used to manufacture clean energy systems are a source of concern and misunderstanding. Steve would probably spin it this way:

> The long-term growth of clean energy is unsustainable. Solar panels and batteries are built with specialized raw materials, and there are simply not enough. Solar cells are already driving up the price of silver and straining its supply chain. The batteries in electric vehicles and grid storage will require more lithium than all known reserves, and these batteries rely on cobalt, most of which comes from a small African nation rife with human rights abuses. As we dig deeper and wider for these raw materials (and others), the mining and processing will create the very type of environmental disasters that clean energy is supposed to avoid.

We keep finding oil and we can keep finding lithium (and other materials)
In the early 2000s a terrifying idea was seizing the American consciousness: "Peak oil" was just around the corner. That is, the price of gasoline and other oil products would begin rising until the world ran out of oil. Economies would collapse. Civil unrest was inevitable. But a few years later, all the dire predictions faded into silence. New technologies like hydraulic fracturing and horizontal drilling, popularly known as *fracking*, helped oil reserves to grow more rapidly than consumption. In 2019, known oil reserves were 20% larger than in 2004.[21]

This is often the nature of natural resources. Fears of depletion usually underestimate the human ingenuity to find more. After all, the stone age did not end because we ran out of stones. For instance, in 2010, the United States Geological Survey (USGS) reported 10 million tons of global lithium reserves.[22] Just ten years later, its estimate had grown to 17 million tons.[23] But with the rapid rise of EVs and batteries, a lot of lithium is consumed—57,700 tons in 2019 and growing. Some analysts now predict we could tap out the world's lithium supplies within the century. This is highly unlikely.

Thanks to the surging interest in EVs, the hunt for new lithium deposits and new methods of extraction have become a billion-dollar industry. The USGS estimates there are an additional 60 million tons of lithium that could be added to the known reserves in coming years. This is probably conservative. During Tesla's September 2020 Battery Day event, the company announced plans to source lithium domestically, claiming there "is enough lithium in Nevada [by itself] to electrify the entire US fleet [of Tesla cars]."[24] And if we assume the battery industry can be at least as innovative in stretching deposits as the fracking industry, global supply chains will gain access to the nearly infinite 230 *billion* tons of lithium in sea water.[25]

What about the other key raw materials in clean energy systems? The silicon and aluminum used in solar cells and racks are both exceptionally abundant, making up 28% and 8% of the earth's crust, respectively.[26] Silver helps solar cells collect the electrons that are freed up when light hits the silicon. Its high cost has motivated the industry to drive down silver usage from 400 mg per cell in 2007 to less than 100 mg in 2019, with predictions taking it down to 50 mg by 2029.[27] Despite this reliance, solar uses less than 12% of the global silver produced annually.[28]

Copper is used in every type of electric device—from televisions to electric vehicles to the wiring in your house. About 20 million tons of copper were mined globally in 2019, a fraction of the 2,100 million tons of identified resources.[29] Solar used about 424,000 metric tons or about 2% of global copper production in 2018. As these numbers show, solar and battery systems rely on only a small portion of the world's production of raw materials and have little risk of running out. A report from MIT summed up the myth of raw material scarcity:

> There appear to be no major commodity material constraints for terawatt-scale PV deployment through 2050. [In fact] Military aircraft production in the United States grew by one-to-two *orders of magnitude* between 1939 and 1944, highlighting the tremendous level of growth that is possible for commodity-based goods.[30]

Of course, any natural resource extraction, whether it is coal or lithium, takes a toll on the environment. As I will explain next, this is why it is critical to understand that natural resources in solar and batteries are consumed in a very different fashion than oil, gas, coal and uranium.

Clean energy does not consume raw materials, it just borrows them

Advocates for 20th century energy often equate the extraction of natural gas, coal, and uranium with the mining of raw materials used in solar and batteries. This misleading comparison reminds me of an old witticism: *When it comes to breakfast, the chicken is involved but the pig is committed.* Resources like coal, uranium, and natural gas are like the poor pig—once they are consumed to generate electricity, they are gone forever. In contrast, natural resources such as silver, lithium, cobalt, and silicon are like the lucky chicken—they need only be put in place once, and they generate and store electricity for decades. After the panels or batteries wear out, the raw materials can be recycled and used over and over.

It is pretty simple. Coal, natural gas, and uranium are fuels. Silver and lithium are not. The "fuel" in renewable energy is sunlight or wind.

The price of clean energy is largely immune to changes in raw material costs

Clean energy systems are *far* less sensitive to price changes in their raw materials than traditional fossil fuel power plants are to their fuel costs. Even if the price of lithium, cobalt, or silver skyrockets, it will have only a modest impact on the price of solar and batteries.

Let me explain. The price of coal and natural gas makes up more than two-thirds of the cost of the kilowatt hours they generate.[31] If the price of natural gas triples, the cost of each kilowatt hour it generates increases more than 200%. But lithium is only 10% of the cost of a battery and silver is just 3.5% the cost of a solar panel.[32] If the price of silver triples, the cost of a panel increases only 7%. Analyst firm Bloomberg New Energy Finance explains this for a popular type of lithium-ion battery, NMC 811:

The sensitivity of battery pack prices to commodity prices is much lower than commonly understood. A 50% increase in lithium prices would for instance increase the battery pack price of a nickel-manganese-cobalt (NMC) 811 battery by less than 4%.[33]

Clean energy is not locked into a small set of raw material options

The raw materials that powered our 20th century grid were mostly fuels, and only three proved to be scalable and cost-effective: coal, natural gas, and uranium. But when the fuel is sunlight, a surprisingly diverse set of raw materials can be used for converting and storing it.

Silicon is the dominant raw material today for solar, largely because of its low cost and flexibility. But if silicon somehow becomes expensive or unavailable, we can turn to already commercialized alternatives that use different raw material recipes including cadmium telluride (CdTe) and copper indium gallium selenide (CIGS), or newer and even more widely available raw materials like perovskites. The same goes for batteries. If lithium becomes scarce, there are increasingly viable alternatives such as zinc, sodium, graphene, and entirely different architectures like flow batteries.

To really appreciate clean energy's flexibility with raw materials, look at cobalt. Ever since John Goodenough created the first practical lithium-ion battery 40 years ago, cobalt has been lithium's quiet wingman. This lustrous, silver-gray metal is hard to find, and most of its production comes as a byproduct of mining other metals, like copper. About 140,000 tons were used in 2019 out of known reserves of 7 million tons. Strategy firm McKinsey & Co projects that demand for cobalt will surge to well over 200,000 tons by 2025, half of this being for batteries.[34]

But cobalt has some big problems. In addition to being one of the most expensive raw materials in lithium-ion batteries, about 70% of the world's cobalt comes from the Democratic Republic of Congo (DRC), a small African country with a history of civil war, child labor, and other human rights challenges.[35] Cobalt reserves are so concentrated in the DRC that the country has been referred to as the "Saudi Arabia of the electric vehicle age." Cobalt experiences wild price swings due to the limited number of sources and the political volatility of the DRC.

So how do you deal with a volatile, expensive raw material like this? You change the battery recipe and remove cobalt as an ingredient. It is not easy to do, but it is possible. On Tesla Battery Day in 2020, Elon Musk announced that, after years of research, the company had engineered cobalt out of its future batteries, simplifying its supply chain and reducing its battery costs at the same time. Most of the world's battery manufacturers will follow Tesla, and within a decade or so, cobalt will no longer be an essential ingredient in the world's battery supply.

MYTH: RETIRED PANELS ARE A WASTE NIGHTMARE

Another over-amplified misunderstanding about solar concerns disposing of retired solar panels. The mythical problem goes something like this:

> Solar claims to be "clean energy," but disposing of millions of retired panels will be an environmental catastrophe. Our landfills will be overflowing with panels. Their toxic materials, like cadmium and lead, will leach into our soil and water. Batteries are even worse, given their greater dependence on toxic materials.

This solar myth has an element of truth to it. Without regulations, retired clean energy systems will join phones, laptop computers, and televisions as part of the lightly regulated and growing e-waste disposal problem. Fortunately, recycling technologies are evolving almost as quickly as solar and batteries themselves, and it may well be cheaper to recycle these components in the future than throwing them away.

Even if we threw away all those panels, they would be only a small fraction of our e-waste

Assume for a moment there was zero recycling and millions of panels end up in landfills. Just how bad might this waste problem be? In 2019, the US installed 19.2 gigawatts of solar.[36] The aluminum, glass, and cells in a panel weigh an average 65 metric tons per megawatt.[37] In 30 to 40 years, when it is time to retire these panels, we will have 1,248,000 tons of material to deal with.

For better or worse, even as the growth in solar accelerates, disposing of retired solar arrays will remain a small part of a much larger waste problem. For example, e-waste from televisions, mobile phones, and computers creates 6.9 million tons of waste annually in the US and is growing quickly.[38] And e-waste is just a tiny part of the 292 million tons of municipal waste generated in the US every year.[39] Even if every 2020 solar panel was thrown into a landfill, it would account for less than 20% of US e-waste and far less than 1% of all the country's waste. In contrast, coal plants generate 100 million tons of toxic ash waste each year. A negligible amount of this makes it to landfills. Almost all of it is stored outside the power plants where it was generated, accumulating to an estimated two billion tons.

How does the waste from retired solar plants compare with the construction waste from retired coal, natural gas, and nuclear plants? When you divide the total construction materials by the number of megawatt hours produced over the plant's lifetime, you see that retired solar panels are not much different from other types of electricity generation. By themselves, solar panels leave 3.2 pounds per megawatt hour at retirement, and 5.2 pounds with racking. Retired natural gas plants leave 2.4 pounds of waste per megawatt hour, and coal plants leave 6.2 pounds. Nuclear plants leave only 1.4 pounds, but the special handling for radioactive materials makes each pound 3 to 10 times more expensive to dispose of.

Waste materials from retired power plants
(pounds of construction materials per megawatt hour over plant lifetime)

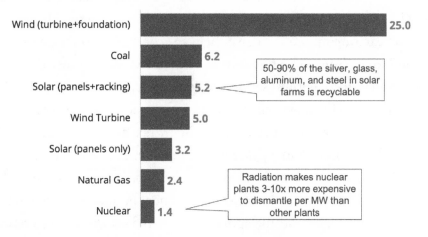

FIGURE 8.4 *Waste from retired solar plants is similar to most other types of power plants. Source: multiple (freeingenergy.com/g106).*

Solar installations are the most recyclable source of energy
The International Renewable Energy Agency is one of several groups that have examined the unique degree of recyclability of solar panels:

> The major components of panels, including glass, aluminum, and copper, can be recovered at cumulative yields greater than 85% by panel mass through a purely mechanical separation.[40]

This is certainly promising, but today's recycling technologies are still developing. The ability to recycle panels does not necessarily make doing so economically compelling. For example, the glass on panels is often coated with films that are expensive to remove. Fortunately, with solar panels lasting 32 years on average, the industry has several more decades to continue improving on the cost-effectiveness of recycling technologies.[41]

Battery recycling, however, is still in its infancy. As electric vehicles become mainstream and their batteries eventually wear out, a lot of innovation is needed to handle the deluge of retired batteries. One exciting option is to give them a second life in our homes and buildings. Consider a 66 kilowatt hour Chevy Bolt battery that has offered years of faithful service. As its capacity declines toward 80%, the car's driving range will shrink proportionally, and the owner may decide to replace it. At that point, the owner has a fully functional 53 kilowatt hour battery *that is fully paid for.* That battery can now be installed in the home and provide the same capacity as *four* Tesla Powerwall residential batteries—for almost no incremental cost!

Even with second-life applications, batteries will need to retire at some point and face the same dilemma as solar panels—recycling versus landfill. Experts explain that separating out lithium, cobalt, and nickel is hard, and achieving the purity required for new batteries is complex and expensive. Will batteries become the next great blight in our landfills? JB Straubel does not think so. His company, Redwood Materials, has raised more than $700 million and is betting that battery recycling will be one of the biggest businesses of the 21st century.[42] If you have not heard of JB, you have certainly heard of the company he co-founded, Tesla. From 2004 through 2019, JB was Tesla's Chief Technology Officer, overseeing virtually every one of the company's breakthroughs, from the industry-changing Model S electric sedan to its revolutionary battery chemistries. JB has commercialized more technology than almost anyone else in the clean energy industry, so it is no small thing that his next big bet is battery recycling. In an interview with Bloomberg, JB explained his remarkable vision:

> There's no real limit to it. There's no degradation that happens to those atoms of lithium or cobalt or nickel. It's one of the coolest things about this—those metals are basically infinitely recyclable. Except for the small amounts that get lost in the recycling process itself, you can basically keep doing that again and again and again, so you can start to imagine a future where you're thinking, 'If we can do this a thousand times, the need for mining new materials starts to dwindle.'[43]

Recycling is not just going to be a big business; it will become a strategic priority for governments. A successful recycling industry will reduce a country's reliance on imported raw materials, not to mention reducing the environmental footprint of mining. This is why governments are pouring money into consortiums like the US Department of Energy's ReCell Center, or the international nonprofit PV CYCLE, both of which are advancing the innovations around lithium-ion recycling.

Solar is one of the least toxic ways to generate electricity

Anti-solarists warn us that solar panels are full of toxic chemicals like cadmium and lead, which can be released into the environment when the panels break or are disposed of improperly. With apologies to Mark Twain, I would like to state that "rumors of death by solar are greatly exaggerated."

Cadmium is often called out as the toxic bad boy of solar. It is a soft, silvery-white metal used to make electroplated steel, color paints, and nuclear reactor control rods. It is also present in the waste ash from coal plants. While cadmium is widely used in other industries, it is used only in about 5% of solar panels, largely because silicon-based solar is cheaper. And even those panels with cadmium use a relatively safer, non-soluble compound called cadmium telluride (CdTe). Animal tests indicate CdTe is 10 times less toxic than aspirin.[44]

The other bad boy is lead. Every year, 11 million tons of lead are produced from mining and recycling, with the vast majority used in lead-acid batteries, like the kind that start your car.[45] In 2018, only 3,900 tons of that lead were used in the global solar panel market, an inconsequential amount.[46] At 14 grams of lead per solar panel, each panel contains half the lead in a shotgun shell or one-thousandth the lead in a car battery.[47]

But it is not enough that other types of energy have larger environmental impacts. Clean energy needs to be held to the highest environmental standards. To this end, the solar industry is following the consumer electronics industry and adopting lead-free solder. The industry predicts that more than half of solar panels will be lead-free by 2030.[48] If governments chose to regulate this more tightly, lead-free solar panels could be a reality at little to no increase in cost.

Panels 2.0

Nearly every solar panel ever built was designed for three big goals: high efficiency, long life, and low cost. But what if we added in a fourth goal—recyclability? It would take only a slight nudge by customers or regulators for this to happen, as the additional manufacturing costs would be offset by the disposal costs of retired panels.

To see the future of solar panel recyclability firsthand, I traveled to Lyon, France, to visit Apollon Solar. Roland Einhaus is the Director of Operations and R&D, and one of the visionaries behind his company's New Industrial Cell Encapsulation (NICE) panels. He is upbeat and enthusiastic, like a friend who cannot wait for you to open a gift he knows you will love. After all, his work has the potential to inspire a transformation in solar energy.

What makes Apollon's new panels revolutionary is their simplicity. They eschew all the plastics, glues, and solders that hold traditional panels together. Instead, cells and copper wires are held in place by the pressure between two big plates of glass and sealed with an environmentally safe, super-long-life adhesive along the glass edges. The design may sound simple, but perfecting and commercializing the NICE technology has taken years.

Recycling a NICE panel requires no chemicals, heat, or mechanical shredding. A simple cut of the seal along the panel's side and the cells and wires fall out, immediately available for reuse or recycling. There are no plastics to separate out, the glass and copper are ready to be used again, and the cells can be recycled or reused as needed. But recyclability is only part of the NICE story. Most solar panels fail for one of two reasons: years of daily heating cycles crack the solder, or the plastic coating yellows with age. NICE's panels avoid both these fatal flaws and can achieve double the lifetime of traditional panels. Most remarkable of all, NICE panels are beautiful. You have to see one to know what I mean, but the iridescence of the cells intertwined with shiny copper wires produces an aesthetic that even non-solar nerds can get excited about.

TRUTH: INTERMITTENCY IS CHALLENGING, FOR NOW

As discussed, intermittency has been solar's Achilles' heel. As the eminently quotable clean energy guru and founder of Bloomberg New Energy Finance, Michael Liebreich, has said, "[even] 2 cents [per kilowatt hour] doesn't make the sun shine at night." The unpredictability of weather is an even larger challenge. Cloudy weather can last for days and cut more than three-quarters of a solar panel's output. The most common criticisms of intermittency go something like this:

> Sunshine is too unpredictable and intermittent to be a source of "baseload" power on the grid. Solar will always be dependent on "dispatchable" power sources like natural gas, coal, and nuclear plants to provide power at night and during cloudy weather. Batteries can shore up solar for a few hours, but they are far too expensive to make solar dependable year-round.

There are legitimate concerns over intermittency today. For many people managing the Big Grid, intermittency looks like an electricity generation problem. In reality, it is an architecture problem that will be revolutionized with technology—innovators are bringing multiple solutions to bear that will minimize and ultimately tackle the challenges of intermittency in the coming years.

People regularly underestimate how new technologies can fix old problems

Countless experts have sat by as new technologies disrupted their industries and rendered their incumbent businesses greatly diminished—or even irrelevant. In 1907, there were 140,300 registered automobiles. The number increased 30 times to 5 million by 1917. It took just one decade for cars to replace horses.[49]

Hydraulic actuators upended the once mighty steam shovel industry. The internet has all but killed printed newspapers, paper magazines, and want ads. Solid-state memory has replaced spinning hard drives in virtually every new computer. These technology revolutions all seem obvious and inevitable in hindsight, but naysayers passionately argued against them every step of the way, even after the revolution was over. As author Upton Sinclair explained, "It is difficult to get a man to understand something when his salary depends upon his not understanding it. "

How can any of us who have come from outside the power industry question the wisdom and experience of the executives, engineers, and academics who design and operate today's grid? In the epilogue to the book, *First on the Moon*, Astronauts Neil Armstrong and Buzz Aldrin wrote, "When an elderly and distinguished scientist tells you that something is impossible, he is almost certainly wrong. The expert can spot all the difficulties but lacks the imagination or vision to see how they may be overcome. The layman's ignorant optimism turns out, in the long run—and often in the short run—to be nearer the truth."

Even Albert Einstein failed to see a future he would help create when he said, "There is not the slightest indication that [nuclear] energy will ever be obtainable. It would mean that the atom would have to be shattered at will."[50]

The economics and business model of solar and batteries are wholly unfamiliar to utility executives. The brilliant and highly competent engineers that design and manage our grids have never dealt with the technical characteristics of solar and batteries. Virtually all experts, including advocates, have failed to predict the degree of performance improvements and price declines of solar and batteries. Assuming the trends continue, can clean energy technologists find ways to address the challenge of intermittency?

Predictability is overrated
Cracking the challenge of intermittency is not dependent on a single, extraordinary breakthrough. Instead, combining existing technologies, all of which are declining in costs, will provide ample options. Here are eight approaches—some of which I've already

touched on—that will blunt or even eliminate the challenges of solar intermittency.

- *Cheap storage.* The cost of both short-term and long-duration storage technologies are declining rapidly. In more and more places, it is now cheaper to generate and *store* solar power across both day and night than it is to buy it from the grid.

- *Vehicle-to-Grid* (V2G). A growing number of EV models can export their electricity back onto the grid. The average EV has enough battery storage to completely power a house for 2 to 3 days. Given the rapid growth of the EV market, these cars will soon become the largest source of electric storage on the grid.

- *Overbuilding solar.* Storage is expensive. Solar is comparably cheaper. As I explain in Chapter 6, "Billion Dollar Disruptions," a growing body of experts are realizing that overbuilding solar to power short winter days is far cheaper than building the batteries to store solar from sunnier summer days. Not only does this slash the cost of a year-round clean energy grid, but the cheap excess summer electricity will also spawn massive new industries.

- *Pairing solar with wind.* These two renewables are surprisingly complementary. The sun shines most during mid-day and in the summer, when days are longer. Wind usually blows most at night and during the winter.[51] When used together, they cut the amount of storage required for daily, seasonal, and weekly intermittency. For example, a 2019 study calculated that battery storage costs of $150/kWh would allow a mix of solar and wind to be cost competitive with the existing grid 95% of the year in many regions.[52] By late 2020, many analysts were reporting batteries had already reached this cost threshold.

- *Country-wide transmission.* Just as one region of the country suffers a rainy week, another has clear skies. Similar variations occur with wind. Investing in cross-country transmission power lines balances out the country's diverse weather patterns and eases the challenge of intermittent renewables.

- *Demand response* (DR). Today, as people turn on their air conditioners and stoves, power plants must increase their output to match the increase in demand. With DR, the cause-and-effect goes in the other direction. As a passing cloud decreases solar output, a smart water heater would pause for a few minutes, reducing the overall load to match the lower solar output. DR is already a multibillion dollar industry.

- *Hydrogen or ammonia.* Until batteries get cheaper, the number of batteries needed to store a week's worth of electricity will be prohibitively expensive. Alternatively, if that electricity is converted into an energy-dense gas like hydrogen or ammonia, it can be stored in relatively inexpensive tanks. Bigger tanks are much cheaper than bigger batteries. These technologies are declining in costs and will soon offer an affordable option for long-duration storage of large volumes of electricity.

- *The grid itself.* Ironically, the existing grid is already solving intermittency for the millions of people with rooftop solar. There are no technical reasons the existing grid cannot solve intermittency at even larger scales. The problem is that the monopoly utilities that own the grid would have to change their business models for this to work.

All of these are on the way to commercial viability today. As the cost of these technologies continues to decline, each one will blossom into its own multibillion dollar opportunity for innovation.

Solar accounted for only 1.7% of the grid in 2019.[53] This low penetration has room to increase 10 or 20 times before existing grid architectures and business models need to consider uncomfortable changes. This gives the technologies above a decade or longer to drive down costs low enough to be a direct replacement for traditional baseload power.

Consider the first computers. They were unreliable, costly, and filled entire rooms. Today, there are billions of computers all around us, and even your smart watch has more computing power than all the world's computers did in the early 1960s. Early mobile phones were once too expensive and limited to replace wired telephony, but two decades later, most people no longer have landlines. If human beings can figure out how to split the atom, we can certainly find ways to keep driving down the cost of batteries.

WHAT ABOUT NUCLEAR?

Why is there a section about nuclear power in a chapter about solar myths? You may find this surprising, but nuclear advocates are some of the most prolific solar myth makers. Yes, nuclear and solar share a common foe in greenhouse gas-spewing fossil fuel plants, but each achieves its low emissions with very different strengths and weaknesses. Over time, as fossil fuels lose ground, and as solar and nuclear continue to innovate, these clean energy systems will inevitably collide with each other. The battle lines are already being drawn as their respective advocates ramp up messaging targeting policy makers and the public.

Complexities and contradictions

If you think splitting an atom is complicated, the politics, operations, and economics of nuclear power can be just as tricky. For instance, if you measure nuclear power by greenhouse gas emissions, volumes of waste, or its ability to provide dependable, baseload power, it is an obvious replacement for fossil fuel power plants. If you measure nuclear by cost, pace of innovation, or potential for catastrophic failures, it is easily the worst way to generate electricity.

But even this is too simple. Nuclear is rife with contradictions. Nuclear proponents boast that uranium is millions of times more energy dense than sunlight. This is true. But safely unleashing the forces inside atoms requires a *multibillion dollar* power plant that takes nearly a decade to build. This price tag makes the kilowatt hours from a new nuclear plant one of the most expensive ways to generate electricity. Nuclear advocates counter that existing US plants, whose construction costs have been paid down over the last 30 to 40 years, are generating electricity at a cost approaching solar. Of course, when a solar plant is paid off after 20 to 30 years, the electricity it generates is effectively free.

Nuclear advocates tell us that their plants are ideal for national security. They can operate for months without refueling, which makes them resilient during crises. Those security benefits are limited, though, since 90% of the fuel used in US nuclear plants comes from foreign nations, half of which is from Kazakhstan, Russia, and Uzbekistan.[54] Solar, on the other hand, requires no fuel. Even if the panels are imported, once they are installed, they will operate for decades with virtually no reliance on outside supply chains, domestic or international.

What about human safety? Nuclear supporters point out nuclear energy has not caused any deaths, at least since Chernobyl, unlike solar, where installers occasionally fall from roofs, causing some injuries and deaths. This comparison is widely made and entirely true—at least if we overlook the much grimmer statistics from indirect deaths from nuclear. The Japanese government reported 2,200 people died as a result of mass evacuations around the Fukushima power plant disaster in 2011.[55]

What about nuclear waste? Proponents point out that if all the electricity you consume in your lifetime were from nuclear power plants, you would leave behind a minuscule amount of waste—about the size of a soda can. Again, this is true. But that can of soda must be put in a steel cylinder, wrapped in a 150-ton cement cask, and protected around the clock until one day in the future when the US identifies a permanent underground storage facility that can safely house the waste for 10,000 years.

As you can start to see, assessing the role of nuclear power as a clean energy solution is as complicated as it is important. The only thing I am certain of is that nuclear power falls well outside the scope of a book focused on rapid innovation, competitive markets, and local decision-making. For these reasons, nuclear power will not play any material role in local energy and, unless things change, nuclear will not play a big role in grid-scale energy either. Let me explain.

The glacial pace of nuclear innovation

Rapid technology innovation and commercialization can take many paths, but the most common and cost-effective routes embrace competitive, commercial markets. This approach is necessarily messy. Failures are inevitable and frequent. Yet the potential for impact and fortune attracts thousands of the world's most talented people, along with hundreds of billions in investments. This market-based approach has driven much of the technology commercialization of the last few decades, including the internet and mobile phones. Pioneers like Elon Musk have also applied it to heavy industries like automobiles and rockets. This is also the approach *Freeing Energy* is championing as a path to accelerating the transition to clean energy.

None of this describes nuclear. The commercialization of nuclear power has never been built on fast-moving, open markets and it never will be—for good reason. Failures can be catastrophic. Extensive regulations and government oversight have made nuclear power far safer than most people realize. But these add substantial costs and delays to nearly every step of the design, construction, and operations of nuclear plants. Nuclear advocates agree that safety regulations are essential, but they argue that building plants in the US has become too burdensome. They point to other parts of the world, like China, where the costs of building new nuclear plants are more reasonable. For instance, a recent report by MIT says that nuclear plants in China are being built at half the cost and half the time of those in the US.[56]

The nuclear "startup" I mentioned in Chapter 3, called NuScale, exemplifies the challenge of US regulations. Their *small modular reactor* (SMR) promises to be safer, smaller, cheaper, and faster to build than traditional nuclear power plants, but NuScale's first scheduled test has been delayed until 2029.[57] Assuming no further delays, it will have taken 22 years and cost more than $1 billion of mostly US government funding.[58] This is a big bet for a single new type of reactor—and, this is just for a test reactor. While there are dozens of other private, early stage nuclear "startups," no other US-based effort is as close to commercialization as NuScale. Assuming the test plant works, it will still take another decade to obtain permits to begin the build out these plants. If NuScale stumbles, it may be another 5 or 10 years before the next closest innovator is ready for tests.

This timescale not only prevents nuclear from participating in mainstream, fast-moving technology markets, it creates a quandary for climate advocates as well. Twenty years until widespread deployment is twice as long as many climate advocates say we have to substantially curtail greenhouse gases. It is difficult to see how the slow pace of the nuclear industry can reduce greenhouse gases in the timeframes that climate advocates argue we need.

Then there is the question of cost. Even the most optimistic outlook for nuclear innovation lacks a clear pathway to cost competitiveness with natural gas, let alone cheaper renewables like solar and wind. "SMRs do make nuclear power more affordable but not necessarily more economically competitive for electric power generation," states a 2018 paper published by the journal, *Proceedings of the National Academy of Sciences*.[59] For nuclear power to expand its role in our clean energy future, governments will need to subsidize the industry at levels several times higher than solar, wind, or fossil fuels, and for many decades to come.

Riding into the sunset

According to his great-grandson, Henry Ford once said, "If I had asked people what they wanted, they would have said faster horses."[60] If you ask the regulators, engineers, and executives in the power industry what they want, they will tell you nuclear power—the faster horse of clean energy. It is easy to see why. From a grid perspective, nuclear is familiar. The economics of nuclear fit comfortably into the centralized asset business model that drives utility profits. Nuclear plants provide steady streams of baseload power, just like coal. In short, nuclear power is a plug-in replacement for our century-old grid architecture. It is fossil fuels without the greenhouse gases. It is the faster horse.

Henry Ford knew that the age of horse-drawn carriages was waning. A faster horse was not the answer. Instead, he created an entirely new approach to transportation. His Model T could travel farther and was more convenient than horses and, in a few short years, the automobile would reshape the way Americans traveled. Today's solar and batteries are the Model T of the clean energy era.

It is easy to mistake this as a race between nuclear and renewables. It is not. Solar, wind, and natural gas are rapidly expanding around the globe, but nuclear is practically standing still. According to the World Nuclear Association, only 50 new reactors are planned across the world in the coming decade. China is the most aggressive nuclear builder with 17 of those reactors.[61] The combined power output of China's planned nuclear plants will be 19 gigawatts. For comparison, in just 2020 alone, China built 48 gigawatts of solar and 72 gigawatts of wind.[62]

Of course, nuclear plants operate around the clock and create three to four times more electricity per gigawatt than solar or wind. But even then, China's renewable additions in 2020 will generate almost twice the electricity per year as all its planned nuclear plants in the coming decade. The US nuclear industry is not even close. It has only added one new reactor in the last 25 years and there is only one new plant currently being built with no others planned.[63]

For nuclear to truly enter the great energy race, it would require more progress in engineering and hard science within the next decade than the industry has seen in a half-century. Costs would need to shrink to a fraction of what they have been, and banks would have to take multibillion dollar bets in the face of cheaper, safer alternatives like solar and batteries. One more Fukushima-type accident could permanently turn public opinion against nuclear, wiping out any of these new investments. Assuming renewables and batteries continue their steady progress on costs, efficiency, and shrinking environmental footprint, nuclear will fall further behind every year.

Of course, a nuclear renaissance is always possible, but it would require extraordinary political will, significant subsidies, and a near-miraculous ability to convince a distrustful public that their fears of radiation are overblown. Without all this support and more, nuclear power will continue its slow decline until it takes its place in history next to fax machines, coal plants . . . and the whimsical notion of faster horses.

(I) (I) (I)

Like Henry Ford's Model T in the early 1900s, solar and batteries are at the dawn of an entirely new era. For a society that was built on horses, transitioning to automobiles was as uncomfortable, and even painful, as it was unstoppable. The transition away from fossil fuels, and possibly even nuclear power, will be no less distressing for many, particularly the incumbents who have provided the fuels for our 20th century energy systems.

As the world reimagines its grids for the 21st century, there will inevitably be a diverse mix of resources. Natural gas has at least a decade of rapid growth ahead of it, perhaps much longer if carbon capture becomes commercially viable. Cleaner sources of energy like geothermal, hydro, wind, and particularly, solar, will play increasingly important roles.

But the most exciting opportunity and the one that so many of us will chose to embrace as individuals, communities, and startups is, of course, local energy. The final two chapters will talk about how innovators, policymakers, investors, and individuals can turn on their own power and make deeply meaningful impacts on our transition to clean energy.

CHAPTER 9

UNLOCKING OUR POWER

"Roads? Where we're going, we don't need any roads."
—Doc Brown, *Back to the Future.*

In late 1993, one of the programmers in my company excitedly waved me into his office. I joined a small group of people who were staring expectantly at a computer. "Watch this," the programmer told us.

He moved the mouse over a button and clicked. I heard the familiar sound of an acoustic modem picking up the telephone line and dialing a number. "This isn't a bulletin board," he said. "We are connecting directly into the internet." Seconds ticked by and eventually a small window popped up, telling us the computer was online.

"Check this out. It is a program called Mosaic. It lets you browse the world wide web."

As he scrolled down, colors, images, fonts of all sizes and types appeared—all of it delivered quickly and fluidly over the internet from a remote computer somewhere far away in cyberspace.

It was the most amazing thing I had ever seen.

With a showman's flourish, the programmer continued, "Keep watching because *this* is what I really want you to see." Another scroll and a click, and up came a new page—and this one had video! It was tiny, maybe two hundred pixels wide. There was no sound. It was grainy. Nonetheless, we all stood there in awe. "What do you think?" he asked me. "Can you see this technology replacing television and VHS tapes?"

"I don't think so," I blurted out. "The bandwidth and processing speed required for passable video are just not feasible. Besides, who wants to sit in front of a computer to watch TV?"

Today, I shudder when I think back to that answer. I should have known better. Ever since grade school, I have been utterly fascinated by the unrelenting progress of technology. I devoured magazine articles and technical specifications about each new generation of microprocessor and memory, jotting down notes and even making graphs. Year after year, experts declared computer technology was reaching its limits—that Moore's Law was running out of steam or that physics prevented more storage density. Undeterred and unconcerned with armchair prognosticators, the industry kept delivering steady improvements. Even today, after decades, the performance and costs of computers continue to improve.

When the programmer suggested a future of streaming video, I knew intellectually that networking bandwidth was based on technologies very similar to computers. I had absolute confidence that decades of steady cost declines lay ahead. Yet despite deep clarity on technology price curves, I was somehow simultaneously in denial about the inevitable outcomes they would create.

My knee-jerk reaction against internet video was a colossal failure of imagination on my part. And not just mine. When Netflix launched its first streaming product fifteen years later, even as its DVD-through-the-mail business model was flourishing, many analysts thought the company was crazy. Needless to say, internet video has become not only ubiquitous (think Netflix, YouTube, and countless other companies), but it has utterly disrupted a once unassailable media industry.

This is exactly what local energy is doing to the power industry. I am determined to avoid another failure of imagination, both for myself and for the world. Because when it comes to clean energy, delaying its adoption a decade or two carries far greater consequences than watching *Game of Thrones* or *The Queen's Gambit* on VHS.

As uncomfortable as it may be to envision something as different as widespread local energy, that reality is coming quickly, and it is reinventing the way we power the world.

Awakening our collective imagination

A decade ago, local-scale solar was more than twice as expensive as buying electricity from the grid (see Figure 9.1). Even with government incentives, high costs limited early buyers to a relatively small customer base of environmentally minded early adopters. It was only in recent years that local solar started to reach cost parity with the electricity rates charged by the Big Grid. This removed cost as a hurdle; but even today, the vast majority of homeowners and local businesses remain on the sidelines, preferring to stick with the familiar Big Grid that served them as children and their parents and grandparents before them.

The most exciting part is that reaching price parity with the grid is just a short stop on local energy's journey of cost declines. Virtually every expert forecasts the same thing—the cost of local energy will fall much, much further for decades. For example, NREL's 2020 analysis shows local-scale solar reaching just 2.3 cents per kilowatt hour in 2050, one-quarter the projected cost of electricity from the Big Grid.

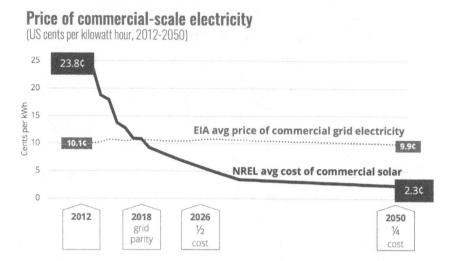

FIGURE 9.1 *The cost of commercial-scale solar power equaled average grid pricing in 2018. By 2050, commercial solar will be one-quarter the price of the grid. Sources: EIA Annual Energy Outlook 2021, NREL ATB 2020 (freeing energy.com/g112).*

What is so remarkable to me is that, despite widespread agreement on the long-term cost declines of local energy and the related lessons learned from computer technology costs, the planners and policy makers in the power industry are as blind to the inevitability of widespread local energy as I was to internet video.

A call for innovators

No one remembers who invented the mainframe computer, but the disruptive impact of personal computers has etched names like Bill Gates and Steve Jobs into every history book. Clean energy will attract far more investment than personal computers ever did. Whose names will be synonymous with this next technology revolution?

This book is written for the ten thousand innovators needed to rapidly make clean, local energy a widespread reality. This is a call to action that goes beyond the usual suspects of technology and VC innovators. Rewiring big parts of the world with local energy will require innovators of every stripe—scientists, engineers, political leaders, advocates, activists, policymakers, accountants, lawyers, professors, teachers, consultants, sales leaders, marketers, IT help desk people, programmers, Fortune 500 executives, economists, think tanks, and government groups. The challenges facing these innovators are great, but the benefits to the world and the personal rewards are even greater.

The rest of this chapter shares the stories of policy, financial, political, and social innovators. The last chapter shares the stories of technology innovators and how they are creating new products and services that are making local energy financially irresistible. Along the way, I will discuss dozens of companies at the vanguard of local energy. For better or worse, even with great management and lots of funding, many of these businesses will struggle and possibly fail. That is the nature of all fast-moving, technology-intensive industries. I mention these companies by name so you can research them to learn more about the segments that interest you. You may even learn more from the companies that fail than the ones that flourish.

But before we dive into the innovators, let me start by painting a picture of where all these innovators will take us. What will the future of local energy look like?

BUILDING ENERGY FREEDOM

Let me take you on a brief tour through your house of the future, which I call *Energy Freedom Home* (EFH).

In many ways, life in your EFH will be the same as it is today. When you flip a light switch or recharge your phone, you will be confident the electricity will always be available. But what goes on behind the outlet and the switch will be entirely different, even for those who are already using today's solar panels.

Electric power will no longer come from the Big Grid or from solar panels attached to the roof. Instead, thanks to a stream of innovations in building-integrated photovoltaics (BIPV), the entire roof will be the panel, along with driveways, sidewalks—even walls and windows. Solar power will become so cheap and ubiquitous that making every surface PV-smart will be as obvious and affordable as getting your house painted.

If you have a need for extra power, sharing electricity between neighbors will be as easy as sharing a cup of sugar, and not much more expensive. The power will flow over the same power lines already in place, with a local electric utility taking a small fee for maintenance and transaction costs.

After the sun goes down, batteries will kick in, seamlessly ensuring an uninterrupted flow of power. These residential batteries will be sold in places like Home Depot and Lowes, right next to the refrigerators—and they will be just as easy to install, upgrade, and take with you when you move.

Your electric car will be more than transportation. It will reduce your home's dependence on the grid, and even generate income as it sits in the garage. Most days, your EV will be charged from your home's solar. But for homeowners who choose to "defect" from the grid altogether, the EVs will occasionally reverse their role, charging at the office or grocery store and later recharging the home's batteries. This will keep your EFH continuously powered, even during extended cloudy weather.

The idea of using an EV to recharge your house will expand well beyond your own car. Just as home delivery companies deliver packages and food today, electricity will be *delivered* to homes in the future. Every Uber and Lyft car, every Amazon, UPS, and pizza delivery vehicle, not to mention a whole new set of purpose-built recharging vehicles, will be available at the touch of an app to come and recharge the home. Fast, wireless charging technologies will allow a hands-free recharge in five to ten minutes. As autonomous, self-driving vehicles become mainstream, the cost to deliver every kind of product, including electricity, will drop substantially.

The EFH will be smarter and make better use of each kilowatt hour. Water heaters, air conditioning, washing machines, and other big appliances will not only be more efficient, but built-in intelligence will determine the least expensive ways for them to run. This will lower the home's costs. And thanks to new business models, it will also help lower the utilities' costs for operating the Big Grid.

All these local energy systems will be interconnected through a vast *energy internet*, allowing electricity to be bought and sold within both neighborhoods and communities, as well as across continents. When a solar roof is not practical or a family cannot finance its own installation, the energy internet will seamlessly deliver clean energy from a neighbor's solar and batteries or from a shared energy project located within the community.

Electricity will no longer come at a cost to the environment. Locally generated electricity will not create greenhouse gases, toxic waste, or airborne pollution. The panels and batteries will be created with clean energy and earth-friendly materials. When it is time to replace them, nearly 100% of their raw materials will be recycled. At some point, mining the raw materials for the world's clean energy systems will slow to a trickle, as sufficient clean energy systems are in place and recycling them is cheaper than extracting new materials.

Smaller versions of the Energy Freedom Home are already changing the lives of hundreds of millions of people in low-income communities in Africa and India, giving their homes electricity for the first time. The technologies and policies that enable the Energy Freedom Home will be applied at much larger scale as well, powering entire neighborhoods, schools, shopping malls, offices, campuses, military bases, small islands, health clinics, hospitals, and police and fire stations.

Every part of this vision is within reach today, using technologies and materials that are already commercially available. Best of all, if enough innovators commit their time and talents, the future of energy will arrive well ahead of schedule.

The inevitability of Energy Freedom Homes and buildings

Local energy is an idea whose time has come. By every measure, it is a better product at a better price. It is the biggest and most immediate way for individuals and local businesses to reduce their greenhouse gas emissions and stop the endless stream of pollution poisoning the planet. Local energy transforms the jobs landscape, from a small number of dirty, dangerous jobs in a handful of hydrocarbon-rich regions into good-paying jobs that are spread across every community and every region of the world. Local energy pulls much of the profits of electricity generation from big, faceless corporations and puts it into the pockets of consumers and communities.

Clean, local energy is also wildly popular. The Pew Research Center found that 84% of Americans support expanding solar.[1] Considering that a Huffington Post poll found that just 76% of people like kittens, solar is one of the most popular ideas ever.[2] Yann Brandt, the widely respected solar executive and author of the *SolarWakeup* newsletter says, "Rooftops are the most valuable and underutilized real estate in the country."

A lot of people agree. A poll from Pew found that 46% of US homeowners are considering installing rooftop solar.[3] According to a study in the *Journal of Economic Geography*, local energy is also a viral idea, with every new solar installation prompting another 0.44 systems in the same area within six months.[4] Online real estate company Zillow found that homes with solar sold for an average of

4.1% more than similar, non-solar homes. That translates to a $9,274 higher resale value for a median-value home.[5]

The interest in local energy will only grow. The potential across the US and the world is enormous. If innovators can surmount the technology and policy hurdles slowing adoption, we can create 500 million Energy Freedom Homes and buildings across the globe and play a huge role in the transition to clean energy.

What AT&T can teach us about betting against innovation

It is easy to assume the status quo will remain such. Most people can only imagine progress in terms of tiny baby steps that stay well within their comfort zones. But history has shown us over and over that big, disruptive changes not only happen, but they can happen with a speed and vigor that changes the world.

If 500 million Energy Freedom Homes sounds audacious, consider another idea that once seemed equally far-fetched. In 1980, AT&T asked consulting firm McKinsey & Company to forecast the number of cell phones that would be in use by 2000. McKinsey predicted 900,000.[6] AT&T took this to heart and limited their bets on the cellular market. AT&T's competitors were far more optimistic. They made huge bets on cellular, and the market took off. AT&T eventually realized it had missed the most important trend in its own industry. So in 1994, AT&T acquired McCaw Cellular for more than $15 billion, one of the largest business deals in US history at the time.[7] And as it turned out, McKinsey was off by three orders of magnitude—there were *740 million* cell phone subscribers in 2000.[8]

Of course, the story of AT&T and mobile phones is only partially about technology. Had it not been for a dramatic shift in policies (not to mention the epic lawsuit the US government won that led to the breakup of AT&T's monopoly), it is likely that our only access to affordable mobile phones, not to mention the internet, would have been limited to reading science fiction books.

Back in the 1970s, AT&T was a regulated monopoly, similar in many ways to today's monopoly electric utilities. Both industries had a big positive impact on American society thanks to the predictability and stability that came from having a single regulated provider. But just like in the power industry, the lack of competition led AT&T to become complacent. The innovation culture of Bell Labs that led to transistors and radar had atrophied. AT&T's unspoken strategy had shifted from changing the world to protecting its own long-term interests.

In another particularly remarkable similarity with the power industry, AT&T began a heavy lobbying and public relations effort to convince legislators that opening the telephony industry to competition was somehow bad for consumers. The company even used the same argument favored by today's electric monopolies: cost shifting harms low-income families. In May 1976, AT&T's CEO, John D. deButts, warned an audience:

> Were the telephone companies deprived entirely of the contribution to common costs that revenues from their more discretionary services provide, they would face the necessity of increasing the average residence customer's bill for basic service as much as 75%.[9]

The harder AT&T fought to maintain its status quo, the harder the world pushed back. A barrage of policy changes and lawsuits from would-be competitors, states, and the federal government resulted in sweeping changes, including the breakup of AT&T and the end of its nationwide monopoly in 1982.

The rest of the story, as they say, is history. Today there are *8 billion* mobile phones in use globally.[10] Economies of volume drove prices down so low that 500 million Africans were able to stop waiting for landlines and connect directly to the world wirelessly.[11] The cost of a long-distance telephone call, ground zero for AT&T's battle to retain its monopoly, is effectively free now. Oh, and fair and open competition also allowed a little thing called the internet to become mainstream.

Similar sweeping policy changes are needed for the electric monopolies but, like AT&T, these will take decades of political lobbying, lawsuits, and state-by-state changes. Fortunately, in the meantime, innovators can press for much smaller, more digestible policy changes across cities, counties, and states that can effectively unlock local power for communities and local businesses, paving a road to Energy Freedom Homes everywhere.

UPGRADING POLICY

Unlocking our power will require more than just technology and business innovations. Policy is the third essential part of making widespread local energy a reality. Most proposals to accelerate the transition to clean energy rely on sweeping, top-down changes in national policy. While dramatic policy changes are always helpful, the US has struggled to make any meaningful federal energy policy that lasts longer than a four-year presidential administration. Local energy stands unique among other paths to clean energy, because many of the policy changes can be made locally, at small scale.

That said, local energy faces a gauntlet of policies that raises costs, adds delays, and too often, renders the Energy Freedom Home impractical or unaffordable. As I discussed in Chapter 7, "Utilities vs the Future," net metering may be the most important policy supporting local energy, but it is under constant assault across multiple states. The details of US electricity policy are staggeringly complex, making the need for savvy policy innovators even more vital to the implementation of EFHs.

Untangling policy

When I first began researching local energy, I was overwhelmed by the byzantine, overlapping regulations and laws governing the grid. To me, US electricity policy was, to borrow Winston Churchill's famous line: "a riddle, wrapped in a mystery, inside an enigma." This complexity was daunting, but it also translated into big opportunities for innovators that could deftly navigate it.

To help me dissect the layers of power policy, I reached out to Steve Kalland, the Executive Director of the North Carolina Clean Energy Technology Center (NCCETC). Steve explained: "The first thing to understand about US electricity policy is that, when you include federal agencies, regional programs, state commissions, self-regulating electric cooperatives, and municipal utilities, there are nearly 3,000 different regulatory forums that have authority over electricity, most of which are unique and interact in complex ways. It gets even more complicated if you are looking at land use and siting issues or air quality issues. At any given location, a distributed energy project is subject to multiple levels of overlapping jurisdictions. This is certainly navigable, but innovators need to understand what they are dealing with."

Let me offer you a 60-second crash course on how all these policies come together to affect local energy. (No need to worry; you will not be tested on this.) It starts locally. Your city, town, or county has a range of policies—building codes, electric codes, permitting, and more. These local policies can include financial incentives as well as zoning restrictions—for example, preventing solar installations on farmland. Even homeowners' associations affect local energy by limiting or banning rooftop solar in the name of aesthetic conformity.

Bigger policies like those governing net metering, community solar, and electricity prices are usually (but not always) state-level. Public utility commissions set policies for the large investor-owned utilities (IOUs), which serve about 70% of the US.[12] The remainder of people in the US are served by smaller utilities called *electric membership cooperatives* (EMCs), *municipal utilities* (munis), and *community choice aggregators* (CCAs). While these utilities can set their own policies on local energy, most are too small to operate their own power generation plants—they typically get power from the IOUs and distribute it to their customers. One byproduct of this arrangement is that many of these small "distribution" utilities have contracts with large-scale electricity providers that frequently limit or outright prevent these small utilities from generating their own power. Overlaying all US power policy is the Federal Energy Regulatory Commission (FERC), which governs interstate electricity transmission and trading. FERC is generally focused on Big Grid issues, but its scope is so broad it can impact local energy, as we learned at the beginning of Chapter 7, where a group of thinly veiled utilities attempted to assert national control over state-level net metering.

It would be challenging enough if the policy confusion ended here, but there is much more. Eighteen states offer *retail choice*, which lets consumers choose between competing retail electricity providers, although this is still highly regulated.

Behind the scenes, there are even more jurisdictions and overlays. If you live in the Northwest or Southeast US, the utilities are *vertically integrated,* meaning a single company owns the entire electricity system, from power plants to the wires connecting your house. The broad scope of these utilities gives them a particularly large influence on all levels of policy.

The rest of the US operates as *deregulated wholesale markets.* These would be more accurately called *slightly-less-regulated* or *restructured* markets, but they do offer some level of competition among power plant owners selling electricity into the regional grids.

Within deregulated markets, there are multiple policy-setting bodies, including balancing authorities, regional transmission organizations (RTOs), and independent system operators (ISOs). Topping it all off, the North American Electric Reliability Corporation (NERC) sets national policies that maintain the stability and reliability of all US grids.

This mind-melting complexity is one of the reason Steve's organization, NCCETC, exists. It maintains the "go to" database for all US electricity policy, called Database of State Incentives for Renewables & Efficiency, or, as everyone refers to it, DSIRE. If you go to the website (dsireusa.org) and type in your zip code, you can get an immediate sense for the power of their data—all the policies and jurisdictions are untangled and streamlined for your location in one simple table.

I asked Steve to sum up the largest policy trends affecting local energy. He said, "Net metering is at the top of the list. Many states are wrestling with new variations, including more sophisticated compensation schemes. Others are trying to get rid of it altogether. We are also seeing a big uptick in policies for distributed storage, advanced grid services, electric vehicles, and grid upgrades that allow for higher penetrations of local energy. Finally, I recommend everyone keep a close eye on FERC Order 2222. While it will take years of wrangling to work out the details, 2222 is specifically aimed at leveling the playing field for distributed energy resources, giving small-scale systems a fair chance to compete in the larger US grid."

Forging a partnership between utilities and local energy

Electric utility monopolies evoke a wide range of public sentiment. They are considered essential by some and called out as malevolent by others. But most people ignore them altogether. A study by consulting firm Accenture found that people think about their utilities roughly 10 minutes a year.[13]

Regardless of public opinion, the fact is that we will need the Big Grid for decades, and we need electric utilities to keep it working. We also cannot shift to 100% solar and wind overnight, any more than we are willing to light our cities with candles or power our factories with steam engines.

But betting everything solely on the Big Grid is far too risky. Politically popular goals like building thousands of miles of new transmission powerlines, reviving the nuclear industry, or building a universal smart-grid are necessary but woefully insufficient. Even if these projects are successful, their advocates acknowledge we will be in the 2030s or 2040s before any broad impact can be realized.

Realistically, many of these "moonshot" goals will fall prey to the old adage that "no one can say yes, and everyone can say no." Even a tiny minority of dissenting stakeholders can tank any one of these mega-projects. The question up for debate is this: how do we safely and justly transition to a future that includes far more clean, local energy, and who can we look to for guidance on navigating this tricky path?

A new generation of policy innovators is emerging. They are pushing the incumbent policy experts into recognizing the inevitable trends of declining technology costs, the improving cost benefits of local energy, and the need to loosen the absolute monopolies that have governed the Big Grid for a century.

Breaking the local energy policy logjam

Navigating these byzantine policies and advocating for increased fairness and access to local energy is tough work. But, thanks to the work of several organizations, substantial progress is being made.

While no advocacy organization can match the resources of the ratepayer-funded utility lobbying, a cornucopia of grassroots organizations is successfully fighting for the energy freedom of homeowners and local businesses. Vote Solar, Local Solar for All, Generation180, the Institute for Local Self-Reliance (ILSR), the Interstate Renewable Energy Council (IREC), and many others are working every day to make local energy accessible to anyone who wants it.[14] Among these organizations, Solar United Neighbors (SUN), founded by Anya Schoolman, has one of the most inspiring stories.

Anya told me that it started when her twelve-year-old son and his friend saw Al Gore's documentary movie about climate change, *An Inconvenient Truth*. They pressed her to get solar for her home. She reached out to a few installers and found the costs and availability made her project prohibitive. Then she had a very big idea—she could combine her purchase with other families to get the prices and service levels she needed. So, they went knocking on their neighbors' doors. She could never have imagined where this would take her.

Two years later, 45 homes in her neighborhood were up and running with their own rooftop solar—and all of her neighbors paid a lot less than the original costs Anya had been quoted.

This would be an amazing story if it ended here. But Anya and the two boys had proved there was a new way to embrace local energy and that community collaboration could lower prices and help push past outdated local regulations. In 2009, Anya created Solar United Neighbors (SUN), and took SUN's approach across the nation. Today, SUN has 100,000 supporters, operates nationally with field programs in 10 states, has helped create 800 jobs, and has enabled 5,800 homeowners to add 46,000 kilowatts of panels to their roofs.[15] The impact on each SUN household is even more impressive. On average, each rooftop solar installation saved $35,000 in electricity spending over 25 years and increased the value of each home by roughly $21,000.[16] I asked Anya what people can do. She said:

> It is easier to get involved than you think. You can start a project, you could write your mayor, you could follow what's going on at your public utility commission. Join one of the many great advocacy groups, including ours. SUN has ways for everyone to get involved at any level of sophistication or time availability. Send a postcard to your legislator, earn a SUN patch for your scouting troop, or organize a solar co-op in your neighborhood.

Hawaii is a postcard from the future of local energy policy

There are a lot of reasons to love Hawaii, particularly if you are an energy policy innovator. With the highest per capita concentration of rooftop solar in the US, Hawaii is pioneering the country's first truly distributed grid. In fact, 60% of the state's clean energy comes from the rooftops of its homes and buildings.[17]

Lani Shinsato of the state's electric utility Hawaiian Electric (HECO) said, "We cannot hope to reach our clean energy goals, especially on land-scarce Oahu, without robust customer participation in generation, storage, and grid services that keep the electric system, reliable, secure, and safe."[18]

At 32 cents per kilowatt hour, Hawaii's electric rates are the highest in the country, causing homeowners and local businesses to flock to rooftop solar.[19] But by 2015, there was so much clean local energy being generated that it threatened the grid's stability. The island's public utility commission thought it had only one option—it stopped letting people put solar on their roofs. Unsurprisingly, the outcry was deafening.

Faced with impossible trade-offs, the island's leaders did something rarely seen in the power industry. They set aside the century-old playbook for grid policy and got innovative. To promote the adoption of local batteries and smart control systems, HECO offers *Smart Export,* which pays only for electricity delivered back to the grid in the early evening, long after solar generation has peaked (this prevents grid overload and reduces the utility's need to burn expensive fossil fuels).

It has worked. During 2020, 80% of Hawaiian rooftop solar installations were paired with a residential battery.[20] The utility also offers *Quick Connect,* which cuts weeks and months of waiting into instant interconnect approvals for new solar installations. Hawaii recently introduced a rate plan specifically for microgrids, only the second state in the US after California to do so.[21] To further drive battery adoption, HECO introduced a program in mid-2021 that pays up to $850 per kilowatt of battery capacity that customers install in their homes. This means customers can receive up to $4,250 toward purchasing a standard five-kilowatt residential battery.[22] Finally, and perhaps most ambitiously, the state is shifting its oversight of HECO from the standard *cost-of-service* rate-making to *performance-based regulation* (PBR). The details of PBR are complex, but it basically means HECO profits are no longer determined solely by how much electricity it sells. Going forward, profits are also determined by a set of key metrics, including the effectiveness of HECO's local energy adoption.[23]

The island's policies are the first of their kind, and there has been no shortage of stumbles and plenty of critics pointing them out. But the state has created a living laboratory for managing high penetrations of local energy. Policy innovators across the world should watch and learn.

POLITICIANS LOVE LOCAL ENERGY

Policy innovators have a powerful and largely untapped benefactor. Politicians may be bitterly divided on grid-scale clean energy, but both parties are surprisingly supportive of clean energy *when it is local,* even if most of them will not stop fighting long enough to realize the difference.

Before clean energy was a political battering ram

It is hard to believe this now, but just over a decade ago, the two major US political parties were surprisingly aligned on clean energy. Both Republicans and Democrats saw it as an essential response to environmental threats, and both saw business and markets as key parts of the solution. The official 2008 Republican Party Platform read: "To reduce emissions in the short run, we will rely upon the power of new technologies . . . we must unleash the power of scientific know-how and competitive markets."[24] The 2008 Democratic Party Platform said, "This challenge [of climate change] is massive but rising to it will also bring new benefits to America. By 2050, global demand for low-carbon energy could create an annual market worth $500 billion. Meeting that demand would open new frontiers for American entrepreneurs and workers."[25]

Sadly, this shared understanding about the importance of clean energy seems like a fanciful notion in today's political landscape. Climate change has become bitterly partisan and clean energy has been sucked into the political vortex along with it. The discord is so intense that both parties are ignoring a topic they agree upon: when clean energy is *local,* it creates jobs in their districts and puts more money directly into their voters' pockets.

An oasis in a hyper-partisan desert

Data from independent surveys as well as actual solar installations reveal local energy is nonpartisan. Voters of all political stripes like it. A 2019 study published in the scientific journal *Nature Energy* analyzed thousands of US households and found that Democrats and Republicans were installing solar at similar rates.[26] More notably, the study uncovered something politicians from both parties should take to heart: Regardless of political affiliation, rooftop solar owners were more likely to vote in both primary and municipal elections compared with their non-solar neighbors—often far more so. Local energy owners are a growing bipartisan political voice.

Electricity is a bedrock foundation for communities, regardless of whether they are wealthy or low-income, rural or urban, Democrat or Republican. Clean, local energy makes every community better. Democrats can embrace lower electric bills that are less likely to be cut off, no unjust delays for power restoration after outages, reduced pollution near families of color, and lots of good-paying local jobs. Republicans can embrace competition, independence from strictly regulated markets, self-reliance, cleaner air and water, and of course, good-paying local jobs. Policy innovators should seize on this rare opportunity for agreement.

REALLY BIG OPPORTUNITIES

To drill more deeply into both the challenges facing local energy and the opportunities for innovators to tackle them, I will now lay out what I call Big Local Energy Opportunities, or BLEOs. Each of these has the potential to unleash a billion-dollar market, and many are already there. Each of these require extraordinary efforts by innovators to reach their full potential. None of the BLEOs require waiting on bureaucratic government mega-policies. Each one can work with a bottom-up, outside-in effort.

By my count, there are over fifty BLEOs. I will step through most of them in this chapter and the next. Taken together, the BLEOs and the innovators that bring them to life will blaze the trail to 500 million Energy Freedom Homes and buildings.

Before we explore these BLEOs, we should take a quick look at the nature of innovation and help dispel some of the myths behind it.

Most innovation is incremental

In 1675, Isaac Newton reflected on his groundbreaking physics insights saying, "If I have seen further, it is by standing on the shoulders of Giants." His statement speaks to the often-misunderstood truth about innovation. It is rarely the product of a eureka moment, nor is it always as simple as a brand-new technology or invention. Instead, innovation emerges through a progression of *innovators*, with each person learning more and pushing their successors a few steps forward.

This pattern of incremental innovation can be seen in the history of internet search engines. Few people remember the early search leaders like Goto.com, Ask Jeeves, AtlaVista, Infoseek, Lycos, Magellan, Inktomi, Excite, and Overture. While all of these have fallen by the wayside, they each played an essential, if overlooked, role in the waves of innovation that ultimately resulted in a search engine that was most preferred by users and by far the most financially successful, Google. Even when the early firms faltered or were acquired, their employees took the lessons and experiences into new companies. Fresh ideas were tested with each new firm. Many failed, but the ones that succeeded became part of subsequent generations of products.

One of Google's most successful features—charging only for ads that are clicked on—was pioneered by a prior firm, Goto.com. Guess who led Goto.com? The same Bill Gross mentioned in Chapter 5, "The Hidden Patterns of Innovation," who started the storage system Energy Vault.[27] The truth is that innovation is far more than companies and technologies. More than anything, it is about the innovators that weave it all together.

The first big opportunities are even better with smart policy

Earlier in the chapter, I explained why local energy is less dependent on sweeping policy change than utility-scale clean energy. But policy still matters. The first set of BLEOs are set to become multibillion dollar businesses on their own, but with the help of some relatively modest policy changes, they can be much bigger still.

Recycling solar and batteries. Clean energy products need to live up to their name and ensure that their environmental impact is as small as possible across their entire lifecycle, particularly when retiring these products. Many project owners will voluntarily recycle their end-of-life systems, but some relatively simple policies can make recycling ubiquitous. For instance, every country in the European Union is subject to the Waste Electrical and Electronic Equipment (WEEE) Directive that requires panels to be collected and recycled after decommissioning. This policy means all the costs and responsibilities for recycling are baked into the upfront price of projects. It guarantees there is a market for recyclers and technology innovators, which in turn attracts investors and innovators. This is one of the best examples of simple policies that can help BLEOs reach scale years or decades sooner than market forces alone can do.

Domestic manufacturing of specialty solar products. China makes more than 70% of the world's solar products. Pandemics, geopolitical tensions, and the occasional plant fire put the clean energy supply chain of every other country at risk. As policymakers play their long, slow, geopolitical poker to try and spur domestic manufacturing of commodity products, innovators can more swiftly rekindle domestic manufacturing by focusing on higher margin, specialized products. Second-order products, from BIPV to integrated local battery systems, will benefit by being geographically closer to their customers, making communications easier and innovation faster. And many specialty products are less price sensitive, avoiding the cost advantage Chinese companies gain with their huge economies of volume.

These specialty segments will emerge on their own, but some simple policies can accelerate them. For example, federal and state governments can require that local-scale projects be built with 100% American-made products. This does not necessarily create a huge market, but it will effectively kick-start the US manufacturing segment, taking an essential first step toward larger economies of volume.

Finding and training millions of people. Transitioning the US to clean energy is going to require nearly a million jobs by 2035, according to the Solar Foundation's National Solar Jobs 2020 Census.[28] Globally, the number is three to five times larger. The biggest bulk of those jobs will be for smaller, local energy projects. Recruiting and training all these people is a Herculean effort (you cannot install solar panels over Zoom). Every installation firm I have spoken with tells me the same story—they cannot find people fast enough. Governments around the world can create incentives and even offer grants to begin training people for these jobs. This opportunity is particularly important in regions like Appalachia, where jobs once focused on fossil fuels can be replaced with installing and maintaining clean energy projects.

100% FOR THE 100%

Low- and middle-income families in the US

Despite its promise, the 100% renewable-powered Energy Freedom Home is beyond the reach of millions of families. Many of these people cannot host their own solar arrays because they rent their homes or live in high-rise buildings. Community solar provides an alternative to rooftops that is increasingly available and increasingly cheaper than electricity from electric utilities. But for a much larger group of people, local energy of any kind remains inaccessible for the most frustrating and tragic reason: they cannot afford it.

Wait, if local energy is cheaper, how can it be unaffordable? The answer is that building a local energy system requires upfront capital, which few families have on hand. Financing this cost is beyond the credit available to many families.

"It is no longer a question of whether the green transition will happen," says Stephanie Speirs, CEO and co-founder of Solstice, a company dedicated to closing that credit gap. "The question is whether that energy system will be just and equitable. Will everyone enjoy the benefits of this green economy, or just a privileged few? Most people don't realize this, but financing rooftop solar or signing up for community solar requires a FICO credit score above 680. Less than half the people in the US have scores that high. An entire segment of the country does not have enough money to save themselves money. This group is disproportionately black and brown, highlighting one of the biggest challenges in energy and environmental equity,"

Solstice is an innovative organization making local energy and energy freedom available to what economists call *low- and moderate-income* (LMI) families. Steph describes her company's customers in more personal terms. For example, there is Joan, a single mom who rents her apartment in New York. Joan told Steph, "I've always wanted to protect the environment. I want the planet to be around for my kids. But I assumed clean energy was just for rich people."

Solstice is changing this. The company analyzed nearly a million lines of credit data and discovered an important and overlooked pattern: people tend to pay their electricity bills no matter what, regardless of their credit score. Solstice created a more accurate and predictive model it calls the EnergyScore, which helps lenders profitably serve thousands of families that would have otherwise never had access to cleaner, cheaper, local energy.

Companies like Solstice are true win-win-win's, promising to help families, help the environment, and provide solid investor returns. Venture capitalist Dave Kirkpatrick of SJF Ventures described why his firm was part of $60 million in funding of another LMI-focused local energy startup, PosiGen.[29]

> SJF has been investing in solar innovators for 15 years. We have looked for models that can scale up residential solar and efficiency to the broader population. PosiGen is the only multi-state company that has developed a model of bundling efficiency retrofits with modular solar PV installations with a 20-year flat lease to provide immediate savings.

There are many others. An organization called GRID Alternatives combines job training and rooftop solar installations targeted at low-income households and communities. The Coalition for Community Solar Access works to expand customer choice and access to solar for all Americans through community solar. Groundswell develops community-oriented solar projects that include houses of worship and community centers.

The combined effects of organizations like these, continually declining installation costs, and enlightened policymaking is working. The income gap for solar is closing, although slowly. A 2021 report from Lawrence Berkeley National Laboratory shows that the median income of households that install rooftop is 20% lower than it was in 2010, from $140,000 down to $113,000.[30] While still well above the average income of $74,000 for owner-occupied homes, the report shows the gap is closing, not getting worse.

Even as innovators are bringing local energy into LMI communities, most utilities are doubling down on their cost-shifting rhetoric, painting LMI families as victims of wealthier solar rooftop owners. Imagine the benefits to come as policy innovators pressure utilities to shift their budgets from lobbying that stops local energy to helping LMI communities expand their access to local energy.

The essential point here is that bridging the credit gap and pursuing energy equity is not just morally essential, it is another big local energy opportunity for innovators of all kinds, from community advocates and policymakers to software programmers and financial analysts. Think about it. Five gigawatts of local energy were installed in the US in 2020—that is roughly 400,000 homes and thousands of businesses.[31] This is an enormous and growing market. As the barriers for people like Joan are torn down, the size of the market will literally double.

Climbing the energy ladder in Africa

Off-grid trade association GOGLA reports almost 100 different organizations are manufacturing solar lighting and phone charging solutions. COVID-19 slowed sales by 35% in the first half of 2020, but the industry will likely surpass US$500 million for the year.[32] This is just the start. A new generation of innovations and innovators are emerging to offer solutions that address *productive use*—tools that allow families and local businesses to generate revenue, sparking a desperately needed virtuous cycle of economic development.

Samir Ibrahim is the CEO and co-founder of SunCulture, the first company in Africa to offer an affordable solar-powered irrigation product.[33] We first met while visiting a smallholder farm in rural Kenya owned by one of his customers. Samir is one of the most remarkable people I met while researching this book. He was recognized by *Forbes* magazine as a "30 Under 30 in Energy."[34] He grew up in the US but moved to Kenya in 2012 to launch his company. A small tattoo on Samir's forearm reads "0609 16.31, 3911 34.08."

He explains, "It's the coordinates of the port of Zanzibar where my family entered East Africa in 1850 as traders."

Samir's family was originally from India but built their lives in Africa. Each of his parents grew up there before moving to Canada where they eventually met. They later moved to Florida, where they ran a dry cleaner for 24 years. The Ibrahim family history inspired Samir to think like a global citizen. He spent much of his university years traveling the world to places like London, Argentina, Peru, India, and China. After graduation he decided to focus the first part of his career on building infrastructure in Africa. It was on this path that he learned about the deep challenges faced by poor African farmers. He had found his mission.

Two-thirds of the African workforce are farmers and about 60% of all the unused farmable land in the world is in Africa. These people are the best hope for the world's growing food scarcity. Sadly, today, these farmers are some of the poorest people in the world. Despite the fact that Africa sits on the world's largest known aquifer, the lack of pumps leaves the farmers entirely dependent on two rainy seasons each year to feed their families. With the growing climate disruption, droughts and floods are wreaking havoc on their crops.

Samir and the SunCulture team set out to find a solution. Their first product was a solar-powered irrigation system. Even with a low-price tag of $5,000, it was outside the budget of most of the farmers they were trying to help. Yet their innovations have been transformative. Through deep customer research and creative technical design, they lowered the price 90%—just above $500 to irrigate a small farm. SunCulture then applied its innovation superpowers to the financial aspect of its offering, allowing farmers to pay for the system over time, using a *pay-as-you-go* solution. SunCulture is now expanding its offerings to include lights, televisions, and other devices.

Solar-powered irrigation can not only help African farmers grow crops year-round, but the predictable and controllable source of water is letting them grow higher-value crops like strawberries and tomatoes. Affordable local energy solutions in the low-income regions of the world will drive unprecedented economic development, leading to improved education, health, and human rights. This may prove to be local energy's greatest impact in the world. But this will also be an enormous business opportunity. Electrifying the homes and communities of the 770 million people who are currently off-grid will draw in investors and create hundreds of thousands of jobs. Even if each person in these low-income regions of the world uses only a quarter of the electricity used by the average American, a brand-new $300 billion annual market will emerge.[35] And with productive use solutions like SunCulture's becoming increasingly available, helping the world's poor is one of the biggest local energy opportunities of the coming decades.

Awareness may be the most important tool to grow local energy

There is one particularly provocative question Silicon Valley investors like to pose to startups that reveals the disruptive potential of their ideas: "What important truth do you believe in that nearly everyone fails to see?"

If you have read this far in the book, I hope you are starting to agree with me that the inevitability of local energy is just such a truth. Yet, even though every data point and every trend point to its widespread adoption, most people still struggle to accept it. Just as I could not see the inevitability of streaming video in 1993, many people today have a hard time wrapping their heads around an electricity system that operates differently than the one they have used their entire lives.

Take, for instance, a story I heard about a parent who offered to donate a rooftop solar array to his daughter's school. Not only was it a sizeable cash gift, the savings on the school's electric bills would make the long-term impact three to four times higher. The school's principal turned it down. He said he had too many other projects and that no one in the school really understood how solar worked. When pressed, he finally admitted that he was skeptical. It sounded too good to be true. He said if it was really such a great deal, there would be solar on every school.

This story reminded me about the skepticism the Wright brothers faced in the years following their successful flight at Kitty Hawk. Despite photos and numerous witnesses, few people who had not actually seen them soaring through the air believed they had flown. When they returned home to Dayton, Ohio, to continue refining their invention, even people who had known the brothers their whole lives were dismissive.

During the first few years, the brothers intentionally avoided publicity to limit copycats as they perfected their invention and filed patents. Even then, the Wright brothers were flying their plane every few days in a field a few miles outside Dayton that was easily visible along a commuter train route into the city. They invited anyone to come and see for themselves, including the local paper.

In his book *The Wright Brothers,* author David McCullough explains the skepticism: "James Cox, [the] publisher of the *Dayton Daily News*, remembered reports coming 'to our office that the airship had been in the air over the Huffman Prairie [5 miles outside Dayton] . . . but our news staff would not believe the stories. Nor did they ever take the pains to go out to see.' Nor did Cox."[36]

Sometimes, the most obvious ideas are the most difficult to see. And the bigger the idea, the harder it can be to accept it.

Why did the Wright brothers succeed when dozens of other well-funded and more famous efforts continued to fail? Wilbur and Orville did the hard work of understanding the fundamentals of flight and wove together a range of innovations that no one before them had ever tried. Successful local energy innovators will do the same.

The last chapter of the book explores some of the technology and product opportunities for these innovators. By understanding the fundamentals and artfully applying different innovations, I am hopeful that a new generation of Wrights and Edisons can make Energy Freedom Homes and buildings a global reality.

CHAPTER 10

POWERED BY INNOVATORS

"We are still in the Model T era of the solar industry. You can have any kind of solar panel you want, as long as it is rectangular and black," jokes Dr. Ben Damiani, borrowing from Henry Ford's famous quote about his cars only being available in black.

Ben is a veteran solar scientist whose innovations have helped shape the global solar industry.[1] I first met him during my early book research. He helped me understand the challenges of turning scientific innovations into mass-produced products. While the industry's feast-to-famine swings have left many solar veterans a bit cynical, calling it the *solar coaster* or *solar circus,* Ben remains a very positive exception. He is confident that solar is on its way to offering homeowners and local businesses true energy independence and self-reliance. He believes that solar can untether people from giant corporations and governments. But he is also quick to point out that there are still enormous technology and policy gaps keeping us from realizing solar's potential.

BUILT FOR LOCAL

One problem, Ben points out, is that almost all the industry's manufacturing energies are focused on only one aspect of the solar business, and it is not the one that empowers consumers or necessarily suits their needs. "Nearly all the innovations in solar panels are focused on the needs of giant utility-scale installations. While this is an important market, these products don't address many of the unique needs for small-scale solar," Ben says. "It is like driving the kids to school and picking up groceries in an 18-wheeler semi-truck."

With a potential market of 500 million Energy Freedom Homes, the opportunity for innovators to deliver *built-for-local* solar and battery products is nothing short of enormous. There are crucial differences between utility- and local-scale solar products. Here are a few examples of areas that are crying out for innovators and new technology.

Seeking sunshine

Shading is the industry term for anything that blocks direct sunlight from reaching the cells inside a panel. This is a bigger problem than you may think. All the cells in a solar panel are connected in series, making it vulnerable to the so-called *Christmas light problem*. Just like old-fashioned strings of light, the failure of one bulb makes the entire string go dark. If even a portion of a single cell inside a solar panel gets shaded, panel power output can drop as much as 33%, or even 100%. The panels in an array are also connected in series, so whatever impacts a panel can impact the entire array.

Shading can affect utility-scale installations, but the problem is manageable. Giant solar farms have to deal with accumulated dirt, called *soiling*, and occasional interventions from Mother Nature, called *bird poop* (this is really the industry's term, and it is a real problem). Fortunately, for these large installations, all this shading can be resolved with something as simple as regular cleaning.

Shading is a far larger problem for smaller systems that you might find on rooftops, where trees, chimneys, vents, and nearby structures may block a portion of panels. Any of these obstructions can cause an entire system's output to plunge. Shading may seem like a mundane problem, but it is worth billions and billions of dollars to the industry. Fortunately, an entire segment, its very own BLEO, has emerged to adapt utility-scale solar panels for local-scale installations.

Two leaders in this market, Enphase and SolarEdge, make electronic boxes that are wired underneath each panel. These devices isolate the shaded panels and prevent them from shutting down the entire array (these technologies also prevent short circuits and dangerous overheating).

How big a problem is this? These two companies alone had a combined market cap over $30 billion (in May 2021), and they have tapped into only one or two percent of the potential market for local energy. Despite their value, these products are like Band-Aids that do little to fix the underlying shortcomings of traditional panels. They also make designs more complex, increase the time required to install wiring and racks, and often cost as much or more than the panels themselves.

Ben thinks there is a better way. "Why not build the electronics directly into the panel? Or better yet, just make the cells a little smarter so we avoid the Christmas light problem altogether?" He argues that both approaches already exist today in real, commercially available products, but many are relatively new and premium-priced.

Heavy, rectangular panels are a bad fit for roofs

Utility-scale solar loves large rectangles. They can be mounted closely together in long, straight rows. The problem is that many home roofs are trapezoid in shape. Rectangular panels result in big, jagged edges between rows. This is unsightly and leaves big portions of the roof without PV. The more promising solution is solar shingles, which I will discuss shortly.

Standard panels are also heavy. A typical rooftop installation uses 25 panels that weigh about half a ton. Many roofs cannot support this much weight. Ironically, the solar cells that produce the electricity are only a few dozen pounds of the overall weight; the rest of it is frames and glass. Early innovators are making panels with lighter components that not only reduce expensive raw materials, but makes installing them safer, faster, and less strenuous. This means more roofs can have solar—and they are cheaper for everyone.

Traditional solar panels duplicate labor and time

All modern solar panels are designed to be attached to a rack. For utility-scale solar, panels are bolted to metal racks secured to the ground. Rooftop solar requires racks as well, but these are only a few inches high and screw directly into the roof. Millions of local energy installations are built this way.

But if you step back and think about it, this approach is kind of silly, especially for new houses. Rooftop racks add material costs and design time, and make installation longer and more complex. Solar panels are already waterproof, and most will outlast standard asphalt shingles. Why are we drilling holes in watertight roofs to add a layer of watertight panels? Why are we sending up two sets of crews, one to install the roof and another to install solar panels on top of it? Why not just use the solar panels *as* the roof?

It turns out this is exactly what innovators are starting to offer and it is called *Building Integrated Photovoltaics*. BIPV is going to be another large BLEO. It reduces materials, it makes the installers far more productive, and it will last longer than traditional roofs. BIPV products eschew big metal frames, so they are much lighter and blend into the roof.

The market is young, and Tesla has garnered the most attention with its solar shingles, but Tesla is not alone. ArteZanos, CertainTeed, Luma Solar, and other companies offer BIPV solutions as well. The biggest provider in early 2021 is a company called GAF Energy, a subsidiary of 130-year-old roofing giant, Standard Industries. GAF Energy's CEO, Martin DeBono, has the confident and focused presence of a decorated former US Navy officer. He served on the Ohio-class ballistic missile submarine, the USS *Georgia*. Martin told me, "You gain a new relationship with energy when you're underwater for 80 or 90 days, and your life depends on a power plant that sits just a few dozen feet away."

GAF Energy's sister company, GAF, partners with thousands of independent roofing companies across the US. This provides his company with a direct line to installers who have decades of experience putting things on roofs. He says, "If you look at rooftop solar installations from 1976, they don't look much different from the solar installations of 2020. The industry has fallen into a rut." Martin is ready to change the way consumers think about home solar installations. "Powering a home with what is, basically, a flatscreen television imported from Asia is yesterday's thinking."

The technologies for BIPV exist today—but weaving all the functions into a cost-competitive product is far from easy. Martin explains, "It is hard to integrate a roof, a water barrier, a fire barrier, and a power plant in one structure that will last for 25 years, but that is what we are doing." The benefits are substantial. "We're eliminating duplicated visits to the customer's home. We're eliminating duplicated trips up and down the roof. We're eliminating duplicate design processes and safety inspections. We're eliminating the homeowner having to deal with duplicate contractors."

When I asked how far BIPV can go, he answered, "I have no doubt integrated solar roofs will take over. There's literally no doubt in my mind. This type of solar power will become the only type for the same reason that people no longer take their horse and buggy to the office. A solar panel is a 40-year-old relic that is long overdue for an upgrade; an upgrade we want to make available to the whole world." Borrowing a term from the computer industry, BIPV is the *killer app* for local energy. It cuts installations costs, solves the weight problem, gets a pass from homeowner's associations, and promises to follow the same cost declines as its bulky solar panel cousins.

"I'm giving it all she's got, captain!"

As fans of the original *Star Trek* series will recall, it seemed like every few episodes, the Starship *Enterprise* would face some grave danger and Captain Kirk would order more power from the ship's engines so they could escape. Inevitably, Chief Engineer Scotty would complain, with a strong Highlands brogue that somehow survived into the 23rd century, that there was not any more power available. Then, just as inevitably, he would find it anyway.

Our Energy Freedom Homes face a similar, but more down-to-earth challenge. Conventional thinkers worry that some homes cannot fit enough solar panels on their roofs to meet all their power needs. Data from Google's Project Sunroof supports this. The data says only 80% of homes can support rooftop solar.[2] The percent of homes whose roofs can host enough solar to power all the home's electricity needs is likely much smaller. Even with BIPV making use of every square foot of roof space, some Energy Freedom Homes will need more power.

With some creative thinking, Scotty found a way to solve the *Enterprise's* power shortage every single time. Innovators are doing the same for our Energy Freedom Homes and buildings. The first step is the simplest and often the cheapest—reduce the power required (and doing so without telling Captain Kirk to shut down life support). Energy efficiency is already a proven market on its own, with annual spending in the US approaching $10 billion a year.[3] It may not be as sexy as some new BIPV product, but is almost always the best place to start, even if the home or building has plenty of solar power.

Efficiency options are plentiful, well established, and will generally pay for themselves quickly: Switch to LED lights and energy efficient appliances. Add insulation around windows and doors, or replace them with newer, more efficient versions. Seal air ducts. Trade-out old heating and air conditioning systems for heat pumps. Install a smart thermostat. Opportunities for innovators are just as diverse, from creating new products and crafting pay-what-you-save financial programs to advocating for creative policies and incentives that make it easier for people to afford upfront investments. But efficiency is just the first step in cutting electricity consumption.

Think about how many more rooftop solar installations could meet all electricity needs if those homes could slash their electricity use by even larger amounts—maybe a third, or even half? According to the US Energy Information Administration, more than 50% of US residential electricity is used for heating and cooling.[4] Geothermal heating and cooling systems can close that gap a lot. Pioneering company Dandelion uses specialized heat pumps connected to pipes buried outside a home. This system uses safe heat-transfer fluids to balance seasonal hot and cold air with stable temperatures available underground. Geothermal heating and cooling products are already an annual $2 billion BLEO in the US alone ($10 billion globally).[5]

If the Energy Freedom Home needs more power, a host of options are emerging, several of which have the potential to become BLEOs on their own. Innovators across the world are pioneering a variety of paints, coatings, encapsulations, and novel modules that transform ordinary elements of our homes and buildings into solar-power generators. Solar rooftops will be joined by solar driveways, sidewalks, walls, windows, garage doors, carports, fences, awnings, patios, and decks. Not all these surfaces will get as much direct sunlight as rooftops, but the declining costs of PV-enabling materials will make these ideas cost-effective and mainstream, nonetheless.

Solar may be the clear leader in locally generated energy, but as I shared in Chapter 3, "The Rise of Local Energy," there are a growing number of small-scale generation technologies vying for a piece of the local energy pie, particularly for larger commercial and institutional projects. Several high-profile startups are working on small, modular nuclear reactors (SMRs). Distributed hydro taps into the 15,000 miles of artificial waterways and the 3.5 million miles of rivers in the US.[6] Distributed wind is a great fit for industrial parks and campuses. A 2020 report from segment leader One Energy found that ultimately, 20% of large commercial and industrial facilities in the US can lower electricity costs and increase resiliency by installing their own, on-site wind turbines.[7]

Fuel cells are another early but growing BLEO. The industry leader, Bloom Energy, generated almost $800 million in 2020 revenue by itself. These systems use catalysts to extract energy from natural gas and generate electricity more efficiently (and cleanly) than combustion-based generators. And as we have already seen, fuel cells can also use super-clean hydrogen, which has no emissions other than water.

As with battery innovation, fuel cell innovators divide their attention between transportation and stationary applications and similarly, both markets benefit. The costs for fuel cells are declining steadily, opening this technology to more customers and markets. In mid-2021, the US Department of Energy launched a sweeping program to lower the cost of generating clean *green hydrogen* four times, from $4 per kilogram down to $1, with the goal of making it cost-competitive with hydrogen created with natural gas.[8]

Powering through the night

Every evening as the sun sets, solar rooftop owners watch their energy freedom fade and their homes and buildings once again become dependent on the Big Grid. Choices about how our electricity is generated shift back to monopoly providers.

Of course, it does not have to be this way. As the plunging cost of batteries continues to follow the pace set by solar, it is becoming easy and affordable to store and reuse solar through the night. The industry calls this *stationary storage*, or *residential storage,* but I prefer to call it *local batteries*. While most of the battery innovation today is aimed at electric vehicles, which have somewhat different requirements from local batteries, there are significant synergies between those markets. Battery R&D has been on a tear in the early 2020s, and it has been fueled by some eye-popping investments.

Battery chemistry pioneers Sila Nanotechnologies and QuantumScape have each raised about $1 billion to commercialize solid-state batteries that promise safer, lighter, and hopefully cheaper electric storage. Solid Power, another early leader, has raised several hundred million. Swedish battery manufacturing giant Northvolt has raised $6 billion to build factories across Europe.[9] Reinforcing the idea from the last chapter that innovation comes from innovators moving among companies, the two founders of Northvolt, Peter Carlsson and Paolo Cerruti, met while working at Tesla.

For local energy innovators, all this investment means lower component prices and more opportunities for innovation in second-order integrations. Tesla's Powerwall remains the most prominent product in the local battery segment as of early 2021, but competitors, both large and small, abound. Early pioneers like Sonnen, SimpliPhi, Blue Planet Energy, and Outback Power Technologies have been joined by larger industrial companies including LG Energy, Panasonic, and Generac. Manufacturing batteries is a modern-day gold rush.

New companies are entering the market every month. Residential-focused battery companies in Asia and Europe are proliferating, and many will compete in the US. New battery chemistries will make these products safer, longer-lived, and far cheaper than they are today. Components will be integrated and combined in ways unimaginable today. Software to optimize and manage batteries will emerge as a major new competitive battlefront. The market for local batteries is still new, but over the long term it will be one of the largest Big Local Energy Opportunities.

Powering through bad weather

When the sky is full of clouds, the power output of solar panels can drop 75% to 90%. A few days of this and even large local batteries will be drained. Once again, the Energy Freedom Home finds itself tethered to and dependent upon the Big Grid.

Extended bad weather is the single biggest obstacle to freeing energy. The industry refers to the technologies addressing this problem as *long-duration storage*—storing electricity for days and weeks rather than hours. There are already commercial solutions available, but they either rely on fossil fuels or are too expensive to compete with the Big Grid. Form Energy, an early, visible leader founded by five battery rockstars[10], has raised more than $300 million in its quest to build ultra-cheap, long-duration batteries.[11] The company claims it can reach $20 per kilowatt hour using its novel iron-air chemistry.[12]

What excites me so much about this segment is the sheer range of innovations that can be created. Will these systems use inexpensive versions of traditional battery cells or something entirely different, like small-scale flow batteries or even thermal or mechanical systems? Perhaps the declining costs of hydrogen electrolysis, storage, and fuel cells will result in local-scale, long-duration storage that can also be used to heat homes at the same time. As we saw with the Stone Edge microgrid in Chapter 3, commercially available hydrogen technologies can be used to store electricity (P2G2P) for much longer periods of time by simply increasing the size of the containers.

Of course, the Big Grid is not going away, and this "backup" for our Energy Freedom Home could be offered at a cost-competitive rate by today's electric utilities using their existing power plants. For neighborhoods, campuses, and towns, the most cost-effective long-duration storage may be at community-scale. Or, as I will get to later, perhaps the best solution to long stretches of cloudy weather will already be parked in our garage.

These are some of the foundational components and integrations of local energy. As largely first- and second-order technologies, they are creating a proliferation of third- and fourth-order BLEOs of their own, which I will cover in the rest of the chapter. I will start with these higher-order businesses by examining them from the decidedly less sexy but absolutely essential perspective of the businesses that actually build the Energy Freedom Homes and buildings.

MAKING THE PRICE IRRESISTIBLE

If there is any single factor delaying our progress toward 500 million Energy Freedom Homes and buildings, it is the cost of local energy. Fortunately, the cost is now generally below the electricity rates on the Big Grid, but cost parity is not nearly enough. For EFHs to be a widespread reality, the price advantage of local energy will have to be so large that it becomes irresistible.

The price of components like solar panels needs to continue declining, but they account only for 15%–20% of local energy installation costs.[13] The biggest part of the price tags are labor and soft costs, which make up about 60% of most projects. Getting the price of local energy down to a half, or even a quarter of Big Grid rates, as the National Renewable Energy Laboratory projects, depends heavily on a group of companies that deserve a lot more attention than they get.

There are more than 5,200 local energy installers in the US.[14] These companies make up their own BLEO but, make no mistake, this is a tough business. It is knocking on doors and climbing on roofs. It is juggling banks, hourly staff, complex software, sales funnels, and just about everything else. Even after the pandemic shrunk the industry a bit, at the end of 2020, it still employed 125,000 people.[15] These installation companies and their people are the vanguard of the local energy industry. To understand how these innovators are tackling the biggest piece of the local energy cost pie, in late 2020, I put on my mask and toured a marquee commercial-scale project with the CEO of one such company.

Every rooftop is a power plant

Michael Chanin is a man on a mission, and it does not take long before he shares it with you. "The built environment of the future will incorporate renewable power on every structure that can support it," he told me soon after we met. Michael is the founder and CEO of Cherry Street Energy (CSE), a firm that builds, owns, and operates commercial-scale local energy projects. With a graduate degree in philosophy from University of Cambridge followed by several years in investment banking including powerhouse Goldman Sachs, Michael is eclectic and defies stereotypes. He is also one of the most compelling evangelists for local energy I have met.

Sustainability is the top priority for CSE's customers, but costs are the biggest factor that determines their speed of adoption. Michael told me, "There are three areas we really have to nail." He explained, "The first is pushing down hardware costs like solar panels. The second is creating smart, cost-efficient ways to finance the projects. And the third is to make our teams as effective as possible. It is hard to overstate the importance of this last area. The more we streamline our processes, from creating proposals to managing completed installations, ultimately improves our customers' experience, saves them money, and lets us deliver high-quality projects sooner."

Streamlining designs and proposals

Companies like CSE always start with a customer proposal. Historically, this required climbing on a roof, pulling out a measuring tape, drafting layouts, designing electrical connections, analyzing sunlight patterns, looking up local utility rates, creating big spreadsheets, and writing long, customized proposal documents. Fortunately, over the last few years, an entire BLEO segment has emerged that offers tools—primarily software—that streamline everything, shaving days of time down to hours or even minutes. "Our industry is still at the beginning of a long overdue software revolution, but it is finally arriving, and we are excited to be at the cutting edge of it," Michael observes.

One example is the design and proposal process. Software companies are automating every step, from using satellite images for roof design to instantly calculating the lifetime cost savings of a proposed project. A company called Aurora Solar became the first *unicorn* in this segment, achieving a valuation over $1 billion for its $250 million fundraiser in the spring of 2021. Several other companies are vying for this market or segments within it, including Energy Toolbase, Folsom Labs (which develops a widely used software tool called Helioscope), and, as you will recall from Chapter 7, Open Solar.

Maintenance and operations

Solar requires less maintenance than any other type of energy generation, but for commercial-scale projects like CSEs, the costs still add up. "The software tools for managing, diagnosing, and billing commercial local energy projects are literally five years behind every other segment of the energy industry," says Michael. He goes on, "It is hard to believe in this digital age, but some of my competitors still don't know they have a component failure until a customer calls them."

Monitoring and maintenance are high-growth segments that are ripe for innovation. Solar-Log, Wattch, AlsoEnergy, and other startups are combining Internet of Things (IoT) hardware and cloud software to gather real-time data across multiple systems and vendors to offer real-time analysis, alerts, and data for billing.[16] The next generation of these tools will use AI to predict failures before they happen, incorporate battery storage and EV charging, and coordinate truck rolls to multiple locations in a day.

Back when I was in venture capital and we were assessing where market segments would evolve, we asked ourselves: is it a feature or a product? Over time, I believe monitoring will become a feature of broader offerings. It will converge with other software segments like battery-management systems, microgrid controllers, billing tools, and many others to form a new BLEO. The resulting segment will be something akin to a local energy operating system.

Software is just one part of the much bigger *operations and maintenance* (O&M) segment. Amicus O&M, Miller Bros Solar, Cypress Creek, and SunSystem Technology are examples of organizations that put boots on the ground to maintain local-scale energy projects.

Analyst firm Wood Mackenzie predicts the global solar operations and maintenance segment will be $9.4 billion by 2025— and that does not even include residential-scale installations.[17]

Building local energy like LEGO bricks

Today, nearly every single solar rooftop, solar+battery system, and microgrid is designed and built one at a time. It makes me think of medieval monks manually copying books by hand before Guttenberg invented the printing press. Fortunately, this is starting to change, and a new BLEO around modular and pre-built systems is emerging.

At the residential scale, Tesla and GAF Energy are both pioneering pre-designed, fixed-size solar and battery options. Rather than each system being unique, Tesla's solar arrays, a different product from its solar shingles, come in only four sizes, simplifying the electrical system design and add-on parts.

Larger, commercial-scale microgrids are going modular as well. I asked Ryan Goodman, CEO of Scale Microgrid Solutions about his company's focus on modularity. His reply:

> Traditional microgrids are all custom. Every component is being integrated for the first time. This means higher operating costs, more engineering, and longer lead times. We are changing the industry with modular, pre-engineered solutions . . . like LEGO bricks. This has several benefits. First, we can install an entire system in six to nine months, twice as fast as custom microgrids. This allows us to compete with traditional backup generators, which are often installed on accelerated timelines. Second, with expensive overhead reduced, we can do the smaller projects profitably that our competitors can't.[18]

This is an active segment with a lot of funding flowing into it. Ryan's company, Scale, raised $300 million in 2020. GreenStruxure, another microgrid developer, raised $500 million in 2021. Other microgrid developers include startup BoxPower and a subsidiary of utility-giant Southern Company called PowerSecure.

Modular systems, project management, and O&M software are just three examples of the innovations that installers and operators are embracing to drive down the cost of local energy. Other BLEOs will focus on improving inverters, racking, sales and marketing software, and cybersecurity. Financing is another area crying out for innovation. I will discuss it in more detail later.

Now that we have reviewed how we can cost-effectively build Local Energy Homes and buildings, let us look at just how much they will transform our century-old grid.

UNLEASHING THE ENERGY INTERNET

On October 29, 1969, just three months after Neil Armstrong took the first steps on the moon, researchers at UCLA in Los Angeles telephoned their counterparts 350 miles north at the Stanford Research Institute (now SRI International) in Palo Alto. They wanted to talk live as they attempted a remote login using their new system, ARPANET, the precursor to the modern internet.[19]

A graduate student at UCLA named Charlie Kline typed an "L" and asked an SRI researcher over the telephone, "Did you get the 'L'?"

SRI replied, "Yes."

Kline typed "O" and asked, "Did you get the 'O'?"

Again, SRI said "Yes, we see the 'O'."

Then Kline typed "G."

But it never arrived. The Stanford system had crashed.

An hour later, the Stanford computer had rebooted, and Kline retyped "LOGIN." That time it worked.

The internet's first communication was a perfect reflection of what it has since become. The very first message was "Lo," striking an epic tone for the internet's own Neil Armstrong moment. Of course, the first three characters ever transmitted over the internet started with "Lo" and included the first letter of the next attempt, "L," that together created "LOL," the modern internet's abbreviation for "laugh out loud." The irony is deep.

Regardless, the revolution had begun.

Fifty years later, there are 22 billion devices on the internet.[20] The sum of all human knowledge is accessible with mere keystrokes. Building the internet was the greatest wealth creation event in human history. The world has irrevocably changed.

What is the energy internet?

The evolution of the internet and the electric grid have taken very different paths, but there are more than a few lessons that can be gleaned from the internet to help design the future of the Big Grid.

When asked to design any kind of reliable network, engineers prefer designs with top-down control, like the original telephone network, early mainframe computer networks and, of course, the Big Grid. All these early networks centralized the "brains" and connected them to "nodes" via relatively simple networks. Telephones were connected over two-wire analog circuits to the telephone companies. Mainframes connected to "dumb" terminals. And the grid delivered electricity from giant centralized power plants and substations over simple wires to passive homes and buildings.

The problem with all of these designs is that they are rife with "single points of failure." Take out a central computer, telephone switch, or critical substation and a big part of the system goes down. In the 1950s and 1960s, when the threat of nuclear war had children practicing duck-and-cover under school desks, military planners went in search of communications networks that could survive a nuclear attack. They needed a network that had no single points of failure.

Researchers went to work. A stream of new ideas like packet switching and peer-to-peer communications culminated in ARPANET, which took its first wobbly steps on that fateful day in 1969. The ideas were wildly innovative, and they flew in the face of conventional thinking about top-down designs. In the face of strong skepticism, the network was built, and it worked. In fact, it worked at a scale that its early designers could scarcely have imagined. If I went back in time and told ARPANET's early innovators that all computer, voice, and video networks in the future would run over packet-switched networks, they would laugh me out of the room.

The internet was not the first effort to link computers, and it certainly was not the most sophisticated. So what is it about the internet that allowed it to eclipse the traditional network designs and replace the incumbent telephone and computer networks? There are many reasons, including public domain technology, but when applied to today's electric grid, there are two powerful design points that enabled the modern internet and are informing the coming energy internet.

First, the nodes, or attached devices, are smart. Each new device on the internet has enough "brains" to facilitate bi-directional communications protocols. The rise of mini-computers, personal computers, and now smartphones meant that all the devices that wanted to connect into the internet were already smart enough. Today, all computer and most telephony nodes are smart. But nearly every node on the Big Grid is as dumb as, well, a toaster.

Second, because the internet nodes are smart, they are not dependent on a centralized hub to communicate. They can talk directly to each other. This is the essence of peer-to-peer. Bob Metcalfe, the inventor of Ethernet, the hardware technology that delivers most of the internet's data, famously posited that the value of a network is the square of the number of devices connected to it. When devices are smart and they can talk directly to each other, the network's value grows geometrically as every new device is added. What does this mean? Any individual or small group can try out any kind of innovation they like. They do not need permission and, more important, the centralized hub does not need to change to accommodate those innovations.

Smart nodes and peer-to-peer created the kind of resilient system the 1960s military wanted for surviving a nuclear attack. Who would have guessed an invention created for national security would also end up streaming episodes of *Gilligan's Island*? Of course, the internet and the grid are very different systems. But we can still learn some important lessons by comparing them.

I want to consider the two design points I mentioned above and explain how they are critical to building the future of electricity.

Dumb grid, meet the smart home

Twenty-five years ago, the typical home was digitally dumb. It was not connected to the internet and only a third of households even had a computer.[21] My, how things have changed. Today, the average American home has twenty-five digitally connected devices.[22] But even as all these devices are getting smarter, the electricity that powers them remains just as dumb as it was one hundred years ago.

Fortunately, smart devices are starting to smarten up the dumb power in your home. In 2020, 11% of US households had a smart thermostat.[23] Even this small segment is expected to become a $14.1 billion market globally by 2025.[24] In 2021, there were nearly 100 million smart speakers like Alexa and Google Home installed.[25]

Even though many of these are performing simple power control tasks like turning on lights, we still have a long way to go. Most of the big power hogs like electric water heaters, washers, refrigerators, and dryers operate blindly, blissfully ignorant of ways they could be saving you money or helping the environment.

The first step toward smart power is understanding how your home is using electricity. Curb, Sense, and Smappee offer monitoring systems that report exactly how your electricity is being consumed in your home. For example, how many kilowatt hours is your refrigerator or drier using every month—or right now?

When I installed one of these systems, I was surprised to learn that my electric water heater was consuming 40% of all the electricity in my house. Fixing this with better insulation and a smarter recirculation pump paid for my monitoring system in a few months. Another company, Span, takes the idea further. Span's products transform traditional circuit breakers into intelligent power hubs. Not only will you know how much power your water heater is using, but you can intelligently turn it off when you are at work to lower your electricity bills.

Adding smarts to the electricity in your home is already a BLEO, and this market is just getting started. In the future, all appliances will be smarter, sharing their status and negotiating with each other to optimize how and when they run. The coming convergence toward a commercial-scale local energy operating system that I mentioned in an earlier section is already playing out at the home level. Apple's Siri, Google's Home, and Amazon's Echo are all vying to be the central brain for your house. And when it comes to controlling the power, the competition will be joined by a wider set of companies like Tesla, General Electric, and your local utility, to name a few. Watch this space carefully, because it is going to be big and exciting.

Making money by making the Big Grid better

Even though modern residential solar and battery systems have computers inside them, most are just as dumb as that toaster when it comes to optimizing the power in your home. By integrating the intelligence in these systems within the home and with the Big Grid, smart homes can transform local energy from money-saving to money-making. Smart local energy systems will follow the same path blazed by telephony and computers—the Energy Freedom Home becomes a smart node on the Big Grid.

The first baby steps use the intelligence in the EFH in the service of the Big Grid. Because this works within the existing monopoly laws that limit all buying and selling to one utility, it is not surprising that this is where policy makers and technology companies have innovated first. As I explained in Chapter 6, "Billion Dollar Disruptions," Virtual Power Plants (VPPs) and Demand Response (DR) coordinate the generation and control of electricity from hundreds or thousands of local energy homes and aggregate them to a scale that can be sold onto the Big Grid.

Installer giant Sunrun is running VPPs across several US states, already generating $50 million a year.[26] Tesla has its own VPP solution that will be used by 50,000 homes in Australia. Other companies in this segment include OhmConnect, Virtual Peaker, Swell Energy, AutoGrid, Germany's Next Kraftwerke, and, the UK's Moxia. This segment has also been one of the most active for M&A, with early leaders Viridity Energy, Greensmith Energy, EnerNOC, and Enbala being acquired in the last few years. The challenge right now is not with private innovation but with policy. Most US power markets are just dipping their toes in the water for VPPs, but for local energy innovators, this will certainly evolve into a BLEO.

Liberating prosumers

One of the biggest trends of the last decade is the rise in "gig" workers. One-third of US adults are now engaged in some form of the gig economy, according to the US government.[27] Generating income outside the traditional 9-to-5 job structure provides financial resiliency, and it allows people to be less dependent on big corporations and governments. This kind of independent income is growing quickly, even as many incumbent industries try to stop it.

The taxi industry spent years attacking Uber and Lyft. The hotel industry is still hard at work trying to stop people from renting out their own homes. Despite these efforts, ride-sharing services like Uber and Lyft generated $46 billion globally in 2019[28] and Airbnb generated $38 billion in income for property owners.[29] In the end, the incumbent industries' efforts to stop these disruptions failed. Consumers are voters, and they tend to get their way when it comes to the freedom to save money and take care of their families.

Like taxis and hotels, the incumbent electric monopolies cannot maintain their lock on the market forever. With Americans building nearly a million solar homes each year, simply reducing electric bills will no longer be sufficient.

The clean energy future belongs to *prosumers*—people and local businesses that produce and consume electricity. Prosumers' Energy Freedom Homes and buildings will become part of a vibrant energy internet, each prosumer generating their own power but also buying and selling it within their community. They will also make additional money selling a range of super technical *ancillary services*, like frequency regulation, black start, peak shaving, and reserve capacity to the Big Grid, helping it remain stable and low-cost. Liberating these prosumers requires a new generation of policy, technology, and business models. More importantly, it requires a new wave of innovators to build it and weave all the parts together.

Peer-to-peer electricity (P2P)

As I said earlier, the first step in creating the energy internet is smart nodes—Energy Freedom Homes that actively coordinate power consumption and generation within a home and across the Big Grid. The second, much larger step, is letting EFHs talk to each other directly, peer-to-peer. It is hard to overstate the importance and disruption of this next step. It creates truly competitive markets for electricity, unleashing innovation onto every aspect of our energy systems. It allows small groups to innovate among themselves without a top-down redesign of the Big Grid. It instigates the first real change in the electric monopolies' business model since Samuel Insull invented it a century ago. In short, P2P is rocket fuel for electric innovation.

While P2P is illegal in the US, other parts of the world, like the European Union, are embracing this kind of energy innovation. In 2018, the EU enacted what is known as the Renewable Energy Directive II, formally defining a concept that had been growing for a decade called *community renewable energy projects* or energy communities.[30] According to the European Commission website, "Energy communities organise collective and citizen-driven energy actions that will help pave the way for a clean energy transition, **while moving citizens to the fore.** [my emphasis]"[31] Groups of homeowners and local businesses are allowed to act as a single electric consumer to the grid, trading among themselves however they like. And as the description states, energy communities (a.k.a. local energy) put citizens' interests first, perhaps for the very first time.

In 2020, French law transitioned from allowing pilots into embracing permanent energy communities.[32] There are different kinds of energy communities, ranging in size from 1 kilometer to 20 kilometers and being capped at certain power levels, like 1 megawatt for small types.[33] At the end of 2019, France was actively running more than 100 pilots. Other European countries have similar stories.

But when it comes to P2P policy and technology, Australia leads the world. Power Ledger is a company based in Perth, with customers across Australia as well as Malaysia, Japan, India, and France. The company offers a pioneering *transactive energy* platform—a software and hardware system that facilitates P2P trading. One study found that the company's technology saved its customers an average of US$424 per year, as well as doubling the savings for rooftop solar owners.[34]

I spoke to Dr. Jemma Green, the Executive Chairman and Co-Founder of Power Ledger, and asked her how this works in practice. She explained:

> We see P2P technology like ours as an operating system for distributed energy markets. It allows local renewables to be scaled in the system without causing many of the unintended consequences that we've seen from very blunt price signals of subsidized renewables. For example, if the local utility is facing an expensive substation upgrade, they can change what they offer into the P2P market. We call this a price signal. This results in a simple economic incentive for local homeowners to upgrade their batteries. I like to think of this as "citizen utilities."[35]

Much as the designers of ARPANET never envisioned Facebook or Netflix, the innovation possibilities of P2P platforms like that of Power Ledger are impossible to imagine today. But one fun example Jemma shared with me was their work with the oldest brewery in Australia. The beer maker was able to power its operations using the solar panels located on the homes of its customers. Power Ledger tracked all the electricity usage and enabled the brewery to compensate their customers directly, with beer delivered to their homes. No money changed hands. Jemma said they call it "peer-to-beer."

Electric vehicles will turbocharge the Energy Freedom Home

In Chapter 7, I took a deep dive into the many ways electric vehicles will fundamentally change local energy. Possibilities abound. Utilities can offer discounts or even payments to consumers in exchange for controlling when EVs are charged. Vehicle-to-Grid technology (V2G) can let EVs send their stored electricity back into the grid, creating VPPs. And as Ford's new F-150 Lightning has highlighted, EVs can be used to directly power Energy Freedom Homes with out-of-the-box V2H technology. The market for V2G technology is expected to exceed $17 billion by 2027.[36]

I spoke with Loren McDonald, the CEO of EVAdoption.com, to learn about the innovators that are already commercializing V2G, V2H, and other EV-based local energy technologies he broadly refers to as V2X.

> For most people, the story of EVs is about reducing tailpipe emissions, but the story is so much bigger. EVs are not just battery-powered cars, they are also becoming a central component in how we power our homes, buildings, and even the bigger electric grid. What makes EV battery storage so powerful is the ability to "time shift" energy from solar parking canopies at the office during the day to the home in the evening when electricity demand increases significantly. The batteries in EVs can not only increase the reliability of the grid, but they can provide consumers and businesses backup power and electricity arbitrage opportunities. Companies like Nissan, Ford, Volkswagen, Lucid Motors, and Hyundai are shipping or have announced plans for bi-directional charging, the key feature that unleashes all these opportunities.
>
> The market for integrating electric vehicles into local energy systems is relatively new, but it is growing quickly.

For example, a company called Nuvve is offering V2G technology for businesses that own fleets of vehicles, including electric buses. Fermata Energy allows companies to reduce the demand charges on their electric bills by using their parking lot chargers and the batteries of participating EVs to shrink their building's peak demand. Another company, WeaveGrid, offers technology that makes it easy for utilities and EV owners to coordinate charging.

Connecting and coordinating EVs into Energy Freedom Homes and buildings will be a particularly large BLEO. Even if you do not own an EV, you will be able to easily top-off your home's local battery by touching an app on your phone and having power delivered in the same way you now order a pizza or receive an Amazon package today. The software and hardware technologies for Mobile Batteries to Home (MB2H) will evolve into its own BLEO.

All these trends will accelerate as batteries continue getting cheaper and longer lasting, alleviating any lingering concerns that V2X will unwittingly reduce the lifetime of EVs.

FUNDING THE REVOLUTION

I spent most of my career in the software tech industry. Whether you are a lifer like me or just a casual observer, when most of us think about funding tech companies, we immediately think of venture capital. But in the world of energy, venture capital is just a small part of a much larger capital story. I was reminded of this the first time I met with Amory Lovins, and he told me that software people often stumble in energy because they do not understand capital assets and the complexity of financing them. At the time, I was not entirely sure what he meant. His words of caution prompted me to look deeply at financing. I have learned a lot since then.

Broadly, I learned that financing local energy is going to require a different mix of sources than the traditional startup world's venture capital-only approach. If we are to build 500 million Energy Freedom Homes and buildings, innovators are going to need to tap venture capital as well as project finance and government funding.

Venture capital

In March 2007, the legendary venture capitalist John Doerr gave an emotional TED talk on the dangers posed by climate change titled "Salvation (and Profit) in Greentech." I was in the audience and remember his opening words: "I'm really scared. I don't think we're going to make it." It was a riveting presentation, one that encouraged listeners to get serious about funding the technologies we will need to reduce the impact of a changing global climate. The venture world listened. Tens of billions were invested in a wide range of cleantech and energy tech startups that all sought better technologies for powering our planet.

In a few short years, it all fell apart. As I shared in Chapter 5, "Hidden Patterns of Innovation," many of those investments bet on solar, battery, and fuel technologies that never scaled to commercial success. More than $10 billion in investments were wiped out. In the early 2010s, cleantech became a dirty word, and access to cleantech funding dried up. Even startups I was covering as recently as 2016 were struggling to find venture investors. This was particularly tough because the founders were all aware of one of cleantech's foundational funding truths: whatever money you think you will need, you will need a lot more.

Fortunately, the VC market has changed—dramatically. A 2020 PwC study called *The State of Climate Tech 2020: The next frontier for venture capital* found that early-stage venture funding for climate tech companies had grown 3750% since 2013, from $418 million to

$16.1 billion of investments in 2019.[37] (The term *climate tech* seems to be replacing *cleantech* to describe technologies aimed at helping the environment; the vast majority of these are energy related.)

In recent years, a new breed of venture firm has been created, purpose-built for the longer cycles, deep capital intensity, and scientific risks that are part and parcel of first- and second-order innovations of energy tech startups. SoftBank, the Japanese-based $100 billion fund, has made several big bets on energy tech, including a $100 million investment in Bill Gross's Energy Vault. Energy Impact Partners is a $2 billion group of funds created by a broad coalition of energy utilities that is laser-focused on electricity innovators. In 2020, Amazon announced a $2 billion fund focused on startups tackling climate change.

The giga-fund launches continued into 2021. Generate Capital announced a $2 billion fund targeting "sustainable infrastructure," General Atlantic announced plans for a new $4 billion fund called BeyondNetZero, and private equity giant TPG announced it had raised a $5.4 billion fund called TPG Rise Climate. Even Hollywood A-listers like Robert Downey Jr. and Leonardo DiCaprio are actively investing in climate tech startups.

If there is one firm that is more visible or more focused on energy than all the others, it is Breakthrough Energy Ventures and its $2 billion fund. Founded by Bill Gates, the list of investors in BEV includes business rock stars and billionaires like Michael Bloomberg (Bloomberg News and former mayor of New York City), Richard Branson (Virgin), Jeff Bezos (Amazon), Ray Dalio (Bridgewater Associates), John Doerr (Kleiner Perkins), Reid Hoffman (LinkedIn), Jack Ma (China's Alibaba Group), and Masayoshi Son (SoftBank).

Bill Gates explained his vision in a 2015 paper he posted to his personal blog, where he wrote: "Promising concepts and viable products are separated by a Valley of Death that neither government funding nor conventional investors can bridge completely. A key part of the solution is to attract investors who can afford to be patient, and whose goal is as much to accelerate innovation as it is to turn a profit."[38]

Incubators and accelerators

As climate tech venture capital funds multiply and look for ever-larger opportunities, another type of organization has grown to support cleantech innovators that are still shaping their businesses. Incubators focus on the earliest stages of innovation, when ideas, products, science, or technologies are still being refined into businesses with commercial potential. Accelerators pick up from there and help innovators scale up their early solutions into larger, high-growth companies. Most incubators and accelerators offer a unique value proposition or focus, like a specific area of technology or a unique target market. Some have competitive selection processes that include prizes. Both types of firms bring some combination of resources to the innovators who join them, including cash funding, access to industry expertise, customer introductions, large corporate partners, and shared resources like accounting, legal, manufacturing labs, and even scientific test equipment.

One such firm, Third Derivative, is the creation of RMI and New Energy Nexus, both of which are mentioned earlier in the book. Third Derivative offers access to an impressive breadth of partners, including investors and large corporations, with the aim of accelerating the growth of climate tech innovators and their companies. Another firm, Greentown Labs, is located a few minutes from the Massachusetts Institute of Technology (MIT), just outside Boston. This is where I first met Francisco Morocz of Heila Technologies, the company whose technology is running the Stone Edge microgrid. Greentown Labs, which recently expanded into Houston, Texas, offers innovators long-term physical space, giving them the ability to build and test products onsite. Startups there also have access to world-class test and manufacturing equipment. For an energy nerd like me, touring the facility and seeing all the early-stage companies and products there was like being a kid in a candy story.

Another firm, and one of the best known, is called Powerhouse. Started by Emily Kirsch, a widely respected clean energy leader, the firm began in Oakland, California, a short drive from Silicon Valley. Since then, this combination incubator and accelerator has added its own venture fund that has invested in many companies, including LMI-focused Solstice mentioned in the last chapter. These incubators and accelerators are part of a much longer list that includes Elemental Excelerator, Techstars Energy, Cyclotron Road, Free Electrons, Joules Accelerator, and many more.

You will find an up-to-date list of firms and incubators that are actively investing in energy tech on my site (freeingenergy.com/vc). In mid-2021, I listed more than seventy-five such investors, and the number is growing every quarter. All of them are pursuing the dual-goal, or *double bottom line* of saving the planet and participating in one of the biggest business opportunities in history.

Funding clean energy projects: project finance

There is one thing that makes clean energy completely different than many other technology-enabled businesses. Clean energy systems are infrastructure. They are part of what Michael Chanin calls the *built environment.* These are built entirely upfront and need to last for decades. Changing a built system is expensive and can undermine the slim profits of these projects. Nearly all the costs of parts and labor are incurred before the system is ever turned on.

Few homeowners or local businesses have the cash to pay for these systems upfront, so most projects require construction loans or project financing. Given the thin profits of infrastructure projects, the financing cost of construction loans has a substantial impact on the ultimate profitability of a project. Shaving even a few percentage points of interest can have a tenfold increase in a project's long- term cash flow. This is why Michael listed financing as one of the three areas firms like his need to get right.

Financing any kind of energy infrastructure, including large local energy projects, is surprisingly complex. Each project typically involves several different types of investors, each taking a slice of the *capital stack*, each looking for different risks and guarantees on their returns.

Equity investors take the highest risk in hopes of getting the highest returns. They are the owners of the project. Debt investors, like banks, accept lower returns in exchange for more guarantees. Tax equity investors evolved due to the complex rules around tax incentives, like the ITC for solar, and their unique types of returns. Insurance is often included to protect against risks like hail damage. Some states allow financing to be repaid as part of a property's annual tax payments, giving lenders higher confidence in their returns. This is called *Property Assessed Clean Energy* financing, or PACE. Cities and counties can pile on their own unique incentives as well as fees and restrictions. On top of all this are transaction costs, like paying lawyers and accountants. These costs rise in relation to the complexity of the capital stack.

Mixing and matching all these ingredients is as much art as science. But practitioners that master this can shave their financing costs and make their projects more profitable than their competitors'. Michael believes financing is a core competitive advantage of his business, so he builds on his experience at Goldman Sachs and manages this internally. Other firms chose to outsource the complexity of financing, and this has created a Big Local Energy Opportunity unto itself.

Dave Riess is the CEO of industry pioneer Wunder Capital. His company specializes in developing and financing commercial-scale energy projects. I asked him why building these commercial-scale projects is important and why it is hard. He told me:

> As I was flying into Los Angeles the other day, I was looking out the window at the acres and acres of white commercial rooftops down below. I counted the number of solar systems on those roofs on two hands. Given the prevailing cost of power in the greater Los Angeles area, each one of the tenants in those buildings would save money by purchasing locally generated power from a rooftop system. So why is it that you see so few solar systems when you look out that window?
>
> There are several reasons, but the biggest one has to do with fixed costs. The transaction costs, like legal paperwork and permitting, tend to be the same regardless of project size. This makes larger projects more profitable because they spread these costs out across more kilowatts. Naturally, developers and financiers go after these larger projects, leaving the smaller project opportunities untouched, even though there are far, far more of them.
>
> Wunder Capital and others are streamlining the process and cost of projects of all sizes, making it more profitable to build even smaller projects. Two-thirds of all electricity in the United States is consumed by commercial and industrial customers, so unlocking this growth is a massive economic opportunity and is also vitally important to addressing climate risks.

Wunder Capital has raised over $500 million and is putting it to work helping its clients more easily and more profitably build local energy systems. Other companies that cater to C&I and nonprofit customers include GreenPrint and RE-volv. Financing smaller, residential-scale local energy systems is a more mature market. Companies like Mosaic, Sunlight Financial, Goodleap, and Dividend Finance streamline the financing process and show just how straightforward financing local energy will eventually be for projects of all sizes.

The US government's surprising role in energy tech innovation

Someone once described the scariest sentence in the English language as: "Hello, we are from the government, and we are here to help." For most of my career, I avoided working with or selling to the government. I saw state and national governments as bureaucratic rat's nests I could never hope to understand. My perspective on this changed as I interviewed dozens of startup execs, scientists, and advocacy leaders for this book. I was pleasantly surprised at how many of them got their start with some help from the government.

What the public/private investment mix should look like and how we bridge that Valley of Death is still very much up for debate. For many, the failure of companies like Solyndra, which lost $535 million of government funding, is a clear sign that the government should stay out of financing clean energy technologies. But as discussed earlier in the book, Solyndra wasn't the only company to benefit from that infamous government loan program.

The Department of Energy's Loan Programs Office (LPO) was created as part of President Bush's sweeping Energy Policy Act of 2005 (which also created the tax credits for solar). The LPO has supported $30 billion in loans.[39] Among the early recipients was Tesla, which received $525 million in 2010 and fully repaid it in 2013.[40]

In early 2021, Jigar Shah, one of cleantech's most respected entrepreneurs and investors, left his venture fund, Generate Capital, to become the director of the LPO, which still has $40 billion in funds to loan out. His willingness to take on this role suggests the fund will be reenergized and once again help fund some of the most cutting-edge technologies in the market.

The LPO is just one of many sources of government funding for energy tech. The Department of Energy has dozens of programs, from grants to prize competitions. In 2017, it provided $12 billion to help fund every kind of energy innovation—from fracking drilling techniques to next-generation solar cells.[41] It is not just the DOE. The Department of Agriculture (USDA) invests hundreds of millions to upgrade rural electric systems, install renewables, and establish smart grid improvements.[42] The Department of Defense actively grants money for research and commercialization of clean energy technologies. (For a full list of all the subsidies, research funding, and loan programs, you can read *"Renewable Energy and Energy Efficiency Incentives: A Summary of Federal Programs"* published by the Congressional Resource Service in November 2020.[43])

Several of the companies I have mentioned in the book—Aurora Solar, the first unicorn in the solar design segment; Mosaic, the residential project finance company; EnergySage, the information portal for solar buyers; and Amicus O&M, the solar maintenance cooperative—all built early momentum thanks to grants from the Department of Energy.

In 2020, I interviewed Dr. Becca Jones-Albertus on my podcast. Becca heads the Solar Energy Technologies Office within the DOE and oversees the department's solar initiatives. She explained how government funding acts as a catalyst to bring together scientists, commercial investors, and business leaders. "Right now, we have about 400 active projects . . . [These] are typically led by national laboratories, universities, businesses, nonprofits, [and] state and local governments, so we have a really diverse portfolio doing a lot of exciting things. . . . We want to support entrepreneurs and help them move as quickly as possible . . ." to bring innovations to market and achieve commercial success.[44]

YOU CAN TAKE IT FROM HERE

The Big Grid has done more for society than almost any other creation of humankind. But it was invented over a century ago, and it is sorely out of date. The Big Grid is poisoning our planet, and it has left nearly a billion people in energy poverty. Massive power outages in Puerto Rico and Texas, along with powerline-sparked wildfires across California, are constant reminders that the Big Grid is also frighteningly fragile. Our electricity system is urgently overdue for an upgrade.

While we wait on politicians, giant corporations, and armies of lobbyists to slog through the necessary but torturously slow policy process, the rest of us can take personal action. We can embrace local energy and make an outsize impact on the transition to clean energy today. Innovators in particular have an essential role to play, because the incumbent power industry has long ago forgotten the pioneering spirit of its founders like Thomas Edison and Nikola Tesla. If clean energy is indeed the biggest business opportunity in history, then local energy is the tip of that spear. Local energy is also the clearest and most actionable path toward a just and equitable energy system that puts the interests of families and communities out in front.

Whether you live in an apartment, in the suburbs, in a city skyscraper, in a hut, or on a farm; whether you are a conservative or liberal; whether you are an employee, an entrepreneur, or a Global 1000 executive; regardless of which county, state, or country you live in the world; whether you are wealthy or barely getting by, you can lend your voice and your energy to this movement and make a bigger personal difference than you might imagine.

This is the first time in the history of our century-old energy system when individuals and small groups can take control, make their own choices, and help create a better future. The technology is here. The political resistance is waning. The greatest threat is simply misunderstandings and ambivalence.

It starts with nothing more audacious than sharing what you have learned with another person, or better yet, sharing it with a policymaker. Every conversation, every new idea you plant in someone else helps local energy move forward. Even better, every solar panel, every electric vehicle, every stationary battery you buy or help bring into existence notches up the incredible economies of volume and pushes down the price the next set of pioneers after you will have to pay.

Each of us has an important role in this collective effort to make the future better, not only for us, but for those who will follow us. As Robert F. Kennedy said, "Few will have the greatness to bend history itself; but each of us can work to change a small portion of events, and in the total of all those acts will be written the history of this generation."

ONE MORE THING

I hope you enjoyed this book. Would you do me a favor?

Like all authors, I rely on online reviews to help spread the word. Your opinion and your support are invaluable. Would you consider taking a few moments now to share your assessment of my book at the review site of your choice? Your opinion will help the book marketplace become more transparent and useful to all.

And if you would like to become part of the Freeing Energy community and play an ongoing role in this important vision, I encourage you to follow us online at:

- LinkedIn: linkedin.com/in/billnussey/

- Website: freeingenergy.com

- Twitter: @freeingenergy

- The Freeing Energy newsletter: freeingenergy.com/subscribe/

Thank you very much!

APPENDIX: A PRIMER ON ELECTRICITY

The three most fundamental components of electricity are voltage, current, and resistance. The underlying physics of electricity are quite complex, but a simple analogy using water and pipes is a great way to understand the basics.[1]

VOLTAGE is like the pressure that forces water through the hose. It is measured in units of volts (V).

CURRENT is like the diameter of the hose—the wider it is, the more water will flow through it. Current is measured in amps (I or A).

RESISTANCE is like the dirt, or gunk, that gets stuck in the hose and slows down the water flow. It is measured in ohms (R or Ω).

Electricity is like a water hose

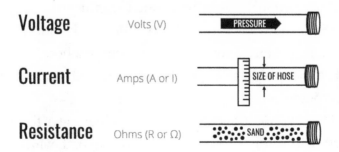

Voltage — Volts (V) — PRESSURE

Current — Amps (A or I) — SIZE OF HOSE

Resistance — Ohms (R or Ω) — SAND

Voltage, current, and resistance are interrelated. This relationship is defined by the equation:

$$V = I \times R$$

According to this equation, voltage is equal to current multiplied by resistance. Going back to our water analogy, if you added sand to a hose and kept the pressure constant, it would reduce the diameter of the hose and result in less water flow.

Voltage = Current × Resistance
(V = I × R)

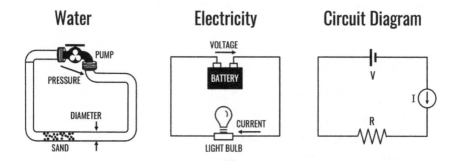

Continuing with the water analogy, we can now discuss how electricity functions in electronics and the grid.

DIRECT CURRENT, or DC, is similar to the normal, single-direction flow (from source to end) of water in a hose. DC electricity is the type used in solar panels, batteries, and inside electronics such as computers, phones, and televisions.

ALTERNATING CURRENT, or AC, is like water flowing back and forth within a hose multiple times per second, which is where our water analogy breaks down. AC is created by electric generators (also called alternators) and is now the global standard for delivering electricity to homes and buildings via the grid.

Alternating Current vs Direct Current

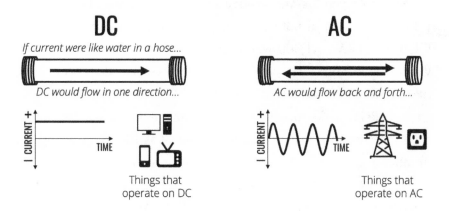

BATTERIES can be thought of as the pump that circulates water through a hose in a closed loop that travels back to the battery. Batteries are capable of generating only DC power. There are several metrics used to describe the capacity of batteries (not all immediately logical). They include amp-hours and kilowatt hours.

TRANSFORMERS are like holding your thumb partially over the end of the hose to make the water spray farther. This is exactly what transformers do for overhead powerlines. The water volume (power) remains the same, but the pressure (voltage) increases as the diameter (current) decreases. This enables electricity to travel farther with fewer losses, because the resistance doesn't impede the electricity when the current is lower. Transformers work only with AC power, which is why AC beat out DC to power our grids more than a century ago.

Electric energy is often confused with electric power, but they are two different things: power measures capacity and energy measures delivery.

POWER is like the volume of water that *is* flowing from the hose at any given instant, given a specific pressure and diameter. It is like gallons per minute. Electric power is measured in watts (W), kilowatts (1 kW = 1,000 watts), or megawatts (1 MW = 1,000,000 watts) depending on the size of the system. Grid-scale power is also measured in gigawatts (1 GW = 1,000,000,000 watts).

ENERGY is like measuring the volume of water that *has* flowed through the hose *over a period of time*. It is like measuring the number of gallons. Electric energy is measured in watt hours (Wh), but most people are more familiar with the measurement on their electric bills: kilowatt hours (1 kWh = 1,000 Wh).

Electric Power vs Energy

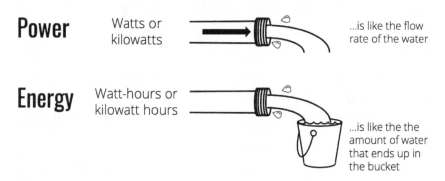

Power   Watts or kilowatts   ...is like the flow rate of the water

Energy   Watt-hours or kilowatt hours   ...is like the the amount of water that ends up in the bucket

A KILOWATT HOUR is the amount of energy that a 1 kilowatt (1 kW = 1,000 W) system will consume or produce in an hour. For example, if you have a 1 kilowatt air conditioner and it runs for an hour, it will use 1 kilowatt hour of electricity.[2]

People that manage the grid, like the electric utilities, prefer to use larger units of energy like megawatt hours (1 MWh = 1,000 kWh = 1,000,000 Wh) or gigawatt hours (1 GWh = 1 million kWh = 1 billion Wh) to measure energy.

ACKNOWLEDGMENTS

This book has been an enormous undertaking. I want to extend my gratitude to the hundreds of people who have contributed their time and skills to bring it to completion. To all of you, I send a huge thank you!

There are two people whose help was so essential that *Freeing Energy* could not have come to fruition without them. The first is the person who has invested more time, more emotion, and more energy than anyone else. She is the book's primary and longest running editor, its bookkeeper and legal organizer, as well as my personal coach, occasional therapist, sounding board, taskmaster, and perpetual supporter—my wife, Melinda. ("Before you keep talking, let me turn on a recorder so we can capture this and transcribe it!")

I also want to thank Jeff Alexander, a seasoned veteran of the book publishing industry and more importantly, my editor, book coach, word master, and friend. Jeff joined this project in late 2017 and has been my collaborator at every step. He wrote the first draft of a few key sections, rewrote a few of my early drafts that really needed help, and has edited every word of the entire book several times. Whenever I ran into a dead end or struggled to find a story, Jeff would help me step back and methodically work through the ideas until we had something we really liked. Melinda and Jeff, this book is really our collective project. Thank you.

One of the biggest challenges of writing this book was finding the people and stories that brought a human face to these complex ideas. Many storytelling ideas came from my longtime friend Sam Easterby. One day back in 2018, Sam suggested how we could get genuine, deeply human stories for *Freeing Energy*. He said, "You should start a podcast! Then you can interview people and get their stories directly." I heard a similar idea from my son Alex and from Michelle Khouri, a friend who went on to start her own podcast production company.

Sam volunteered to be the podcast's producer, a role he has played every two weeks without fail ever since. He helps me find the guests, drafts the discussion outlines, marketing headlines, and pull quotes, and works with the super-talented audio engineers (Peter LoPinto, and most recently, Sean Powers) who have turned our raw "um's" and "ah's" into sharp, succinct, professional-sounding audio productions. So many of the stories and quotes for the book were born from the podcasts and the nearly 70 brilliant and wonderful guests who have shared their time, inspirations, and words with our audience.

As many writing mentors advised me, it takes a relentless commitment to keep pressing forward, week after week. A particularly great piece of advice was to find a writing or accountability buddy. In addition to Melinda, I want to thank another friend, Loren McDonald. As I toiled away at *Freeing Energy*, he was launching what ultimately became his own full-time effort, a website and research firm called EVAdoption. Over the years that we read each other's work and stress-tested provocative ideas, Loren's firm has become a serious and widely respected voice in the electric vehicle industry. He has been a constant companion over email and Zoom these last few years. I am not sure how I could have gotten this far without his nonstop encouragement. Thank you, Loren!

I was fortunate to have several book-writing mentors, but one stood above the rest: Emily Loose, a New York-based veteran book industry editor with many best-selling titles on her resume. Her keen understanding of nonfiction business books like mine was extraordinarily helpful from the very beginning to the final draft of the manuscript. I should add that Emily is a world-class editor. It took me several weeks to work in the suggestions and changes she made, and the book is so much better for it. Emily, you are a rock star, thank you!

When I started this journey, I had few contacts in the industry. James Marlow, whom I have known from our days in the software world and was one of the first of us to jump into clean energy, took me under his wing and opened his Rolodex.

A few months into the book research, I met Joseph Goodman. At the time, he was with RMI, and now has his own climate tech venture fund. Joseph was wonderfully generous with his network, and I met most of the people at RMI through him. When I visited RMI many months later to interview Amory Lovins, I spent an hour walking through the snow and brainstorming with another then-RMI person, Titiaan Palazzi. He later cofounded his own exciting AI energy tech company. He has introduced me to more people than I can count and helped me arrange the tour of Stone Edge microgrid. Thank you, James, Joseph, and Titiaan.

Back when my early enthusiasm for clean energy was looking for an outlet, it was my longtime friend, Mike Hettwer, who called and said, "I know what you need to do. You need to write a book!" That turned out to be one of the most important ideas of my life. Thank you, Mike. I am also grateful for the boundless support of my friend Peter Simms. He has been a writing mentor and was the person who introduced me to Emily Loose, who ultimately introduced me to Jeff Alexander. Thank you, Peter.

It was Ben Damiani, more than anyone, who helped my interest in solar turn into a passion. Ben's love for the science, engineering, and business of clean energy was infectious, and his occasionally crazy and more often brilliant ideas inspired several sections of this book.

I also want to thank Jacqueline Novogratz. She has been endlessly supportive and has opened my eyes to the breadth of good that can come from local energy.

My family has been so supportive and encouraging, there are no words to describe it. I thank you all—Melinda, my grown boys, Alex and Ben, and my mother, Joyce—from the bottom of my heart.

Three people deserve a special shoutout: Jon Long, Ethan Freishtat, and especially Bailey Damiani, helped me carry much of the weight that built the foundation for this book. They each spent hundreds of hours as "interns" helping me gather information, do quality checks, write stories, and perform analysis. Thanks also to Donna Weller and Danny Samuels, who helped with a lot of the hard, slow work of getting a well-researched book to completion. Thank you, all.

Amory Lovins and Jim Rogers were two of my first interviews. They both inspired me, and Amory's vision became the roadmap for much of *Freeing Energy*.

I want to thank many of the brilliant and inspiring people I met at RMI, including Roy Torbert, Sophia Misfud, Megan Kerins, Jon Cretys, Stephen Doig, and Marty Pickett. There were several people while I was still at IBM that strongly influenced my decision to embrace my passion for clean energy, most notably Chris O'Connor, Peter Karns, Chandu Visweswariah, Karl Sandreuter, Roger Premo, and Arvind Krishna.

Thanks to Eric Hartz, Anand Thaker, Sam Easterby, Ben Damiani, Joseph Goodman, Joe Lassiter, Pat Rezza, Marty Hollingsworth,

Karyn Mullins, Samir Ibrahim, Jim Fox, Michael, Barnard, Peter Kelly-Detwiler, Jack Timothy, Tracy Moore, Jon Shieber, and Joyce Choplin for reading early drafts of the book.

During the first year or two of my research, several people spent hours of their time teaching me about the world of clean energy. Even though their names are not in the book's chapters, they were just as helpful as many of the people who are mentioned. These folks include Joe Lassiter (Harvard Business School), Paul Wormser (CEA), George Touloupas (CEA), Yosef Abramowitz (Energiya Global Capital), Ken Dulaney (North Carolina State University FREEDM Center), Deepak Divan (Georgia Tech Center for Distributed Energy), Marc Perez, Tim Lieuwen (Georgia Tech's Strategic Energy Institute), Ashvin Dayal (The Rockefeller Foundation), Autumn Proudlove (North Carolina Clean Energy Technology Center), Chris Porter (National Grid), Baird Brown, Michael Naylor, Salil Pradhan, Don Wood (DFJ), Tim Echols (Georgia Public Service Commission), Martin Hermann (BrightNight), Frank Faller (EDP Renewables), Sam Goldman (d.light), Kareem Dabbagh (VoLo Earth Ventures), Kiran Bhatraju (Arcadia Power), Kristen Brown (Electron), Michael Skelly (Grid United), and Ryan Stoltenberg (Stone Edge microgrid and Helia).

As I made my way into the clean energy community, several tribe makers invited me into their communities and welcomed me into the industry. Thanks to Nico Johnson (SunCast podcast), Jason Jacobs (*My Climate Journey* podcast), Tor "Solar Fred" Valenza, Yann Brandt (Solar WakeUp), and Mike Casey (Tigercomm).

Whether with words, images, or colors, storytelling is truly an art. I am grateful to the many people and their amazing creative skills for helping shape the brand and imagery of Freeing Energy. Thanks to Jon Hutson, Eric Pinckert, Patrizia Schopf, and Alice Goodsmith of Brand Culture for their artful skills in honing the story and message for the book's release. Thanks to Michele DeFilippo and Ronda Rawlins for the interior design and production (1106 Design), Ian Koviak and Alan Dino Hebel for the cover design (theBookDesigners), and Liz Townsend and Laura McDonald for making Freeing Energy stand out visually.

And, last but by no means least, I want to thank the dozens and dozens of people who spent hours with me as I worked to share their stories and visions in this book. Their names are scattered throughout the chapters of this book, so I will not list them all here.

Creating Freeing Energy and writing this book of the same name has been the greatest professional journey of my life. I am deeply grateful to all the people, including those whose names are not in this book or in this acknowledgment, for their support, their inspiration, and the work they do every day to make the world a better place.

ABOUT THE AUTHOR

Bill Nussey has been a tech CEO for most of his career. His first company, which he cofounded in high school, provided graphics software for early, text-based personal computers. His second company, Da Vinci Systems, was started out of his college dorm room and grew to serve millions of users across 45 countries. Later, Bill spent several years at venture capital firm Greylock,  after which, he became the CEO of internet consulting firm iXL, where he helped the company go public, grow to 3,000 people, and achieve $500 million in annualized revenue.

Next, he was CEO of Silverpop, a global leader in cloud-based marketing. After growing to nearly $100 million in revenue, Silverpop was acquired by IBM and became the foundation of IBM's Marketing Cloud. After the acquisition, Bill was promoted to VP Corporate Strategy out of IBM's headquarters in New York. He worked with

IBM's CEO and SVPs, as well as other corporate leaders to help chart the company's strategic direction and roadmap. Bill fell in love with clean energy and left IBM to launch several projects, including a solar science startup, a TED talk, a website, a podcast, and this book. Over his career, Bill's companies have created thousands of jobs and billions of dollars in value.

Bill received a degree in electrical engineering from North Carolina State University and an MBA from Harvard Business School. He holds two patents and has published three books. Bill is an avid reader, an avowed nerd, and he loves movies. He and his wife, Melinda, live in Atlanta, Georgia. They are both involved with several local energy projects serving disadvantaged communities in Africa and the US.

ENDNOTES AND CITATIONS

CHAPTER 1
IN SEARCH OF ENERGY FREEDOM

1. "Hurricane Maria Communications Status Report for Oct. 4." *Federal Communications Commission*. 06 Oct. 2018. Web. <https://www.fcc.gov/document/hurricane-maria-communications-status-report-oct-4>.

2. Mifsud, Ana Sophia, and Roy Torbert. "Schools Stronger than Storms." *RMI*. 27 Sept. 2018. Web. <https://rmi.org/schools-stronger-than-storms/>.

3. Fox, Michelle. "Restoring Puerto Rico's Power is going to be 'overwhelming,' PREPA says." *CNBC*. CNBC, 29 Sept. 2017. Web. <https://www.cnbc.com/2017/09/29/restoring-puerto-ricos-power-is-going-to-be-overwhelming-prepa-says.html>.

4. Puerto Rico, Mayoral Office of Carmen Yulin Cruz. Executive Order 064. 2017.

5. "Resilient Power Puerto Rico." *RPPR*. 2018. Web. <https://prenergytoolkit.com/>.

6. Terrasa-Soler, José Juan, and Daniela Lloveras-Marxuach. "Community Power As Provocation: Local Control For Resilience And Equity." *Scenario Journal*. 2020. Web. <https://scenariojournal.com/article/community-power/>.

7. Kern, Rebecca. "Rooftop Solar Nearly Doubles in Puerto Rico One Year After Maria." *Bloomberg Environment*. Bloomberg Industry

Group News, 20 Sept. 2018. Web. <https://news.bloombergenviron-ment.com/environment-and-energy/rooftop-solar-nearly-doubles-in-puerto-rico-one-year-after-maria>.

8. Vila Biaggi, Ingrid, Cathy Kunkel, and Augustin Irizarry Rivera. *We Want Sun and We Want More: 75% Distributed Renewable Generation in 15 Years in Puerto Rico Is Achievable and Affordable.* Rep. Institute for Energy Economics and Financial Analysis, Mar. 2021. Web. <http://ieefa.org/wp-content/uploads/2021/03/We-Want-Sun-and-We-Want-More_March-2021.pdf>.

9. Pickerel, Kelly. "An Insane Amount of Energy Storage Was Deployed in the United States in Q3 2020." *Solar Power World.* 03 Dec. 2020. Web. <https://www.solarpowerworldonline.com/2020/12/an-insane-amount-of-energy-storage-was-deployed-in-the-united-states-in-q3-2020/>.

10. Fox-Penner, Peter S. *Power after Carbon: Building a Clean, Resilient Grid.* Cambridge, MA: Harvard UP, 2020. Print.
Appreciation and apologies to Peter Fox-Penner and his book, *Power After Carbon,* which introduced the term "Big Grid" and for allowing me to use it in a slightly different way.

11. Constable, George, and Bob Somerville. *A Century of Innovation: Twenty Engineering Achievements That Transformed Our Lives.* Washington, DC: Joseph Henry Press, 2003. Print.

12. "The US Interstate Highway System Cost $548 Billion to Build over 40 Years." *Freeing Energy.* Web. <https://www.freeingenergy.com/m187/>.

13. "The Apollo Moon Missions Cost $152 Billion." *Freeing Energy.* Web. <https://www.freeingenergy.com/m186/>.

14. *MoneyTree Report.* Rep. no. Q4. PWC, 2020. Web. <https://www.pwc.com/us/en/moneytree-report/assets/pwc-moneytree-2020-q4.pdf>.

15. Murtaugh, Dan. "Goldman Sees $16 Trillion Opening as Renewables Pass Oil and Gas." *Bloomberg Green.* Bloomberg, 17 June 2020. Web. <https://www.bloomberg.com/news/articles/2020-06-17/goldman-sees-16-trillion-opening-as-renewables-pass-oil-and-gas>.

16. "Remarks as Delivered by Secretary Granholm at President Biden's Leaders Summit on Climate." Energy.gov. DOE, 23 Apr. 2021. Web. <https://www.energy.gov/articles/remarks-delivered-secretary-granholm-president-bidens-leaders-summit-climate>.

17. Lovins, Amory. "Transcript of "A 40-year Plan for Energy." *TED.* Web. <https://www.ted.com/talks/amory_lovins_a_40_year _plan_for_energy/transcript>.

CHAPTER 2
YOUR POWER IS FAILING

1. Gaffney, Austyn. " 'They Deserve to Be Heard': Sick and Dying Coal Ash Cleanup Workers Fight for Their Lives." *The Guardian.* Guardian News and Media, 17 Aug. 2020. Web. <https://www.theguardian. com/us-news/2020/aug/17/coal-spill-workers-sick-dying-tva>.

2. "EPA Response to Kingston TVA Coal Ash Spill." *EPA.* Web. <https:// www.epa.gov/tn/epa-response-kingston-tva-coal-ash-spill>.

3. *Facility Report.* Rep. EPA, Mar. 2021. Web. <https://enviro.epa. gov/triexplorer/release_fac?p_view=COFA&trilib=TRIQ1&sort=_ VIEW_&sort_fmt=1&state=47&county=47145&chemical=All+che micals&industry=ALL&year=2008&tab_rpt=1&fld=RELLBY&fld =TSFDSP>.

   Kingston Ash Spill. Rep. Environmental Integrity Project, 8 Dec. 2009. Web. <https://web.archive.org/web/20140624160138/http:// www.environmentalintegrity.org/pdf/newsreports/TVA_2008TRI_ Kingston%20Ash%20Spill%202009128.pdf>.

   Releases: Chemical Report. Rep. EPA, Mar. 2021. Web. <https:// enviro.epa.gov/triexplorer/release_chem?p_view=USCH&trilib =TRIQ1&sort=_VIEW_&sort_fmt=1&state=All+states&county=A ll+counties&chemical=007440382&industry=2211&year=2007& tab_rpt=1&fld=RELLBY&fld=TSFDSP>.

4. Bourne, Joel, Jr. "Neglected Threat: Kingston's Toxic Ash Spill Shows the Other Dark Side of Coal." *Environment News.* National Geographic, 19 Feb. 2019. Web. <https://www.nationalgeographic. com/environment/article/coal-other-dark-side-toxic-ash>.

5. Satterfield, Jamie. "On 10th Anniversary of Kingston Coal Ash Spill, Workers Who Went 'through Hell and Back' Honored." *Knoxville News Sentinel.* Knoxville News, 22 Dec. 2018. Web. <https://www.knoxnews.com/story/news/crime/2018/12/22/kingston-coal-ash-spill-workers-10th-anniversary/2333826002/>.

6. This is just one coal waste dam disaster. In 1974, 125 people died and 4,000 were left homeless when a dam failed at a coal mining site in Buffalo Creek, West Virginia. More recently, in 2014, a drainage pipe in North Carolina broke, dumping 27 million gallons of coal slurry into the Dan River.

7. "Frequent Questions about the 2015 Coal Ash Disposal Rule." EPA, 02 Nov. 2018. Web. <https://www.epa.gov/coalash/frequent-questions-about-2015-coal-ash-disposal-rule>.

8. "Mapping the Coal Ash Contamination." *Earthjustice.* 05 Oct. 2020. Web. <https://earthjustice.org/features/map-coal-ash-contaminated-sites>.

9. *Ash Around the World: Surveying the Latest Research and Applications in CCPs.* Rep. American Coal Ash Association, 2018. Web. <http://web.archive.org/web/20181209222134/https://www.acaa-usa.org/Portals/9/Files/PDFs/ASH01-2018.pdf>.

10. "US Coal Ash Waste Could Cover All of New York City under 9 Feet of Ash." *Freeing Energy.* Web. <https://www.freeingenergy.com/m191/>.

11. "Cleaner Power Plants." *Mercury and Air Toxics Standards.* EPA, 23 Oct. 2020. Web. <https://www.epa.gov/mats/cleaner-power-plants>.

12. Vohra, Karn, Alina Vodonos, Joel Schwartz, Eloise A. Marais, Melissa P. Sulprizio, and Loretta J. Mickley. "Global Mortality from Outdoor Fine Particle Pollution Generated by Fossil Fuel Combustion: Results from GEOS-Chem." *Environmental Research* 195 (2021): 110754. Web. <https://www.sciencedirect.com/science/article/abs/pii/S0013935121000487>.

13. "How Much Land Has Been Disturbed by All Surface Mining in the United States?" *Large Coal Mines.* Global Energy Monitor, 30 Apr. 2021. Web. <https://www.gem.wiki/Large_coal_mines

#How_much_land_has_been_disturbed_by_all_surface_min-ing_in_the_United_States.3F>.

14. "Minutes of Meeting of the Board of Directors Tennessee Valley Authority." TVA, 22 Aug. 2019.

15. "Egyptian Pyramids." *Wikipedia*. Wikimedia Foundation. Web. <https://en.wikipedia.org/wiki/Egyptian_pyramids>.

16. "Laurentide Ice Sheet." *Wikipedia*. Wikimedia Foundation. Web. <https://en.wikipedia.org/wiki/Laurentide_Ice_Sheet>.

17. Jacoby, Mitch. "As Nuclear Waste Piles Up, Scientists Seek the Best Long-term Storage Solutions." *C&EN*. 30 Mar. 2020. Web. <https://cen.acs.org/environment/pollution/nuclear-waste-pilesscientists-seek-best/98/i12>.

18. "It's Closing Time: The Huge Bill to Abandon Oilfields Comes Early." *Carbon Tracker Initiative*. 18 June 2020. Web. <https://carbontracker.org/reports/its-closing-time/>.

19. The US Energy Information Administration (EIA) defines biomass as including municipal-waste-to-energy, landfill gas, and biogas. However, the statistic used here is purely wood-based biomass.

20. *Economic Benefits of Increasing Electric Grid Resilience to Weather Outages*. Rep. DOE, Aug. 2013. Web. <https://www.energy.gov/sites/prod/files/2013/08/f2/Grid%20Resiliency%20Report_FINAL.pdf>.

21. Hofmann, Alex. "Defending against Outages: Squirrel Tracker." *American Public Power Association*. 26 July 2017. Web. <https://www.publicpower.org/blog/defending-against-outages-squirrel-tracker>.

22. "Northeast Blackout of 2003." *Wikipedia*. Wikimedia Foundation. Web. <https://en.wikipedia.org/wiki/Northeast_blackout_of_2003#Sequence_of_events>.

23. *Event Notification Report for February 16, 2021*. Rep. NRC, 16 Feb. 2021. Web. <https://www.nrc.gov/reading-rm/doc-collections/event-status/event/2021/20210216en.html>.

24. Hauser, Christine, and Edgar Sandoval. "Death Toll From Texas Winter Storm Continues to Rise." *The New York Times*. 14 July

2021. Web. <https://www.nytimes.com/2021/07/14/us/texas-winter-storm-deaths.html>.

25. "Camp Fire (2018)." *Wikipedia*. Wikimedia Foundation. Web. <https://en.wikipedia.org/wiki/Camp_Fire_(2018)>.

26. "CPUC Orders Deployment of Microgrids and Resiliency Strategies to Support Communities and Infrastructure Threatened by Power Outages." *YubaNet*. California Public Utilities Commission (CPUC), 11 June 2020. Web. <https://yubanet.com/california/cpuc-orders-deployment-of-microgrids-and-resiliency-strategies-to-support-communities-and-infrastructure-threatened-by-power-outages/>.

27. Larson, Aaron. "Expert: 90% of U.S. Population Could Die If a Pulse Event Hits the Power Grid." *POWER Magazine*. 22 July 2015. Web. <https://www.powermag.com/expect-death-if-pulse-event-hits-power-grid/>.

28. Stone, Maddie. "The US Is Finally Heeding Warnings About a Monster Solar Storm." *Gizmodo*. 30 Oct. 2015. Web. <https://gizmodo.com/the-us-is-finally-heeding-warnings-about-a-monster-sola-1739620903>.

29. Maehara, Hiroyuki, Takuya Shibayama, Shota Notsu, Yuta Notsu, Takashi Nagao, Satoshi Kusaba, Satoshi Honda, Daisaku Nogami, and Kazunari Shibata. "Superflares on Solar-type Stars." *Nature* 485.7399 (2012): 478-81. Web. <https://www.nature.com/articles/nature11063>.

30. United States. Cong. Senate. Committee on Homeland Security and Governmental Affairs. 114th Cong., First sess. S. Rept. U.S. Government Publishing Office, 2016. Web. <https://www.govinfo.gov/content/pkg/CHRG-114shrg22225/html/CHRG-114shrg22225.htm>.

31. Woolsey, R. James, and Peter Vincent Pry. "The Growing Threat From an EMP Attack." *The Wall Street Journal*. 12 Aug. 2014. Web. <https://www.wsj.com/articles/james-woolsey-and-peter-vincent-pry-the-growing-threat-from-an-emp-attack-1407885281>.

32. Singh, Divyesh, and Sahil Joshi. "Mega Mumbai Power Outage May Be Result of Cyber Attack, Final Report Awaited." *India

Today. 20 Nov. 2020. Web. <https://www.indiatoday.in/india/story/mumbai-power-outage-malware-attack-1742538-2020-11-20>.

33. Duster, Chandelis. "Energy Secretary Says Adversaries Have Capability of Shutting down US Power Grid." *CNN.* Cable News Network, 06 June 2021. Web. <https://www.cnn.com/2021/06/06/politics/us-power-grid-jennifer-granholm-cnntv/index.html>.

34. "Small-scale Power Grid Attack Could Cause Nationwide Blackout, Study Says." *Fox News.* FOX News Network, 13 Mar. 2014. Web. <https://www.foxnews.com/politics/small-scale-power-grid-attack-could-cause-nationwide-blackout-study-says>.

35. Martin, Christopher. "Rooftop Solar Seen Protecting U.S. Power Grid From Attack." *Renewable Energy World.* 30 Apr. 2013. Web. <https://www.renewableenergyworld.com/solar/rooftop-solar-seen-protecting-u-s-power-grid-from-attack/>.

36. "Access to Electricity." *SDG7: Data and Projections – Analysis.* IEA. Web. <https://www.iea.org/reports/sdg7-data-and-projections/access-to-electricity>.

37. Oglesby, Cameron. "The Navajo Nation Generates a Ton of Power—but 14,000 Homes Don't Have Electricity." Grist. 05 May 2021. Web. <https://grist.org/justice/navajo-nation-electricity-power-covid/>.

38. *Electricity Access in India: Benchmarking Distribution Utilities.* Rep. NITI Aayog, Oct. 2020. Web. <https://smartpowerindia.org/Media/WEB_SPI_Electrification_16.pdf>.

39. *Africa Energy Outlook 2019.* Rep. IEA, Nov. 2019. Web. <https://www.iea.org/reports/africa-energy-outlook-2019>.

40. "Energy for All Program." Grid Alternatives. Web. <https://gridalternatives.org/what-we-do/energy-for-all>.

41. Plumer, Brad. "U.S. Nuclear Comeback Stalls as Two Reactors Are Abandoned." *The New York* Times. 31 July 2017. Web. <https://www.nytimes.com/2017/07/31/climate/nuclear-power-project-canceled-in-south-carolina.html>.

42. Monk, John. "Ex-SCANA CEO Kevin Marsh Pleads Guilty to Conspiracy Tied to Failed Nuclear Project." *The Slate.* 24 Feb. 2021.

Web. <https://www.thestate.com/news/local/crime/article249477970. html>.

43. Dooley, Debbie, and Sara Barczak. "Rewarding Failure: Taxpayers on Hook for $12 Billion Nuclear Boondoggle." *The Hill.* 02 Apr. 2019. Web. <https://thehill.com/opinion/energy-environment/436876-rewarding-failure-taxpayers-on-hook-for-12-billion-nuclear>.

44. "A New Dawn for Nuclear Power?" *The Economist.* 19 May 2001. Web. <https://www.economist.com/leaders/2001/05/17/a-new-dawn-for-nuclear-power>.

45. "The Rise of PCs Caused the Price of Computing to Fall 100 Million times." *Freeing Energy.* Web. <https://www.freeingenergy.com/m192/>.

CHAPTER 3
THE RISE OF LOCAL ENERGY

1. "MicroGrid." *Stone Edge Farm Estate Vineyards & Winery.* 25 Jan. 2019. Web. <https://www.stoneedgefarm.com/microgrid/>.

2. "MicroGrid Tour." *Stone Edge Farm MicroGrid.* Web. <https://sef-microgrid.com/overview/tour/>.

3. "What's in a Megawatt?" *SEIA.* Web. <https://www.seia.org/initiatives/whats-megawatt>.

4. "MicroGrid." *Stone Edge Farm Estate Vineyards & Winery.* 25 Jan. 2019. Web. <https://www.stoneedgefarm.com/microgrid/>.

5. "State Electricity Profiles." *Independent Statistics & Analysis.* Energy Information Administration. Web. <https://www.eia.gov/electricity/state/>.

6. "Puerto Rico Territory Energy Profile." *Puerto Rico Profile.* Energy Information Administration, 19 Sept. 2019. Web. <https://www.eia.gov/state/print.php?sid=RQ>.

7. Nussey, Bill. "Why are millions of people putting solar on their roofs?" The Freeing Energy Podcast, 10 Nov. 2020. Web. <https://www.freeingenergy.com/podcast-vikgram-aggarwal-solar-rooftop-energysave/>.

8. Schneider, Mycle, Antony Froggatt, Friedhelm Meinass, M. V. Ramana, Agnes Stienne, Amory Lovins, Ben Wealer, Tadahiro Katsuta, Christian von Hirschhausen, and Julie Hazemann. *The World Nuclear Industry Status Report 2019*. Rep. Mycle Schneider Consulting, Sept. 2019, p. 18. Web. <https://www.worldnuclear-report.org/IMG/pdf/wnisr2019-v2-hr.pdf>.

9. "Local-scale Solar Produces as Much Energy as 5 Nuclear Plants." *Freeing Energy*. Web. <https://www.freeingenergy.com/m194/>. David Feldman, Matthew Zerling and Robert Margolis, National Renewable Energy Laboratory Q2/Q3 2019 Solar Industry Update, November 12, 2019

10. Shah, Jigar. *Creating Climate Wealth: Unlocking the Impact Economy*. Denver, CO: ICOSA Pub., 2013. Print.

11. Jordan, Phil. *Wages, Benefits, and Change*. Webinar. National Association of State Energy Officials, 6 Apr. 2021. Web. <https://static1.squarespace.com/static/5a98cf80ec4eb7c5cd928c61/t/60772d6c9a200430a1ff75a5/1618423165067/2020+Wage+Report+Presentation-April+6+Webinar_+Final.pdf>.

12. "National Solar Jobs Census 2020." *The Solar Foundation*. 06 May 2021. Web. <https://irecusa.org/programs/solar-jobs-census/>.

13. "Residential solar installations create the most jobs per megawatt." *Freeing Energy*. Web. <https://www.freeingenergy.com/g207/>.

14. "Fastest Growing Occupations: Occupational Outlook Handbook." *Occupational Outlook Handbook*. U.S. Bureau of Labor Statistics, 09 Apr. 2021. Web. <https://www.bls.gov/ooh/fastest-growing.htm>.

15. Farrell, John. "Solar Surprise: Small-Scale Solar a Better Deal than Big." *Institute for Local Self-Reliance*. 08 Mar. 2019. Web. <https://ilsr.org/solar-surprise-small-scale-better-deal/>.

16. Song, Kaihui, Shen Qu, Morteza Taiebat, Sai Liang, and Ming Xu. "Scale, Distribution and Variations of Global Greenhouse Gas Emissions Driven by U.S. Households." *Environment International* 133 (2019): 105137. Web. <https://www.sciencedirect.com/science/article/pii/S0160412019315752>.

17. Nussey, Bill. "Straight Facts on the Environmental Impact of Coal: CO2 Emissions, Pollution, Land, and Water." *Freeing Energy*. 16 Sept. 2020. Web. <https://www.freeingenergy.com/environmental-impact-coal-water-co2-so2-mercury-pollution/>.

18. "Solar and Wind Use 250 Times Less Water than Coal and Nuclear Plants." *Freeing Energy*. 03 Nov. 2020. Web. <https://www.freeingenergy.com/g107/>.

19. "Comparing CO2, SO2, NOx, and methane across coal, natural gas and biomass." *Freeing Energy*. 03 Nov. 2020. Web. <https://www.freeingenergy.com/g105/>.

20. Rand, Joseph, Mark Bolinger, Ryan H. Wiser, Seongeun Jeong, and Bentham Paulos. *Characteristics of Power Plants Seeking Transmission Interconnection As of the End of 2020*. Issue brief. Lawrence Berkeley National Laboratory, May 2021. Web. <https://emp.lbl.gov/publications/queued-characteristics-power-plants>.

21. "Disconnected: The Need for a New Generator Interconnection Policy." Americans for a Clean Energy Grid, 05 Jan. 2021. Web. <https://cleanenergygrid.org/disconnected-the-need-for-new-interconnection-policy/>.

22. "How Much Electricity Is Lost in Electricity Transmission and Distribution in the United States?" *Frequently Asked Questions (FAQs)*. Energy Information Administration, 18 Nov. 2020. Web. <https://www.eia.gov/tools/faqs/faq.php?id=105&t=3>.

23. Fitzgerald, Garrett, James Mandel, Jesse Morris, and Herve Touati. *The Economics of Battery Energy Storage*. Rep. Rocky Mountain Institute, Oct. 2015. Web. <https://rmi.org/wp-content/uploads/2017/03/RMI-TheEconomicsOfBatteryEnergyStorage-FullReport-FINAL.pdf>.

24. Smil, Vaclav. "Electricity: It's Wonderfully Affordable, But It's No Longer Getting Any Cheaper." *IEEE Spectrum*. 29 Jan. 2020. Web. <https://spectrum.ieee.org/energy/policy/electricity-its-wonderfully-affordable-but-its-no-longer-getting-any-cheaper>.

25. Wesoff, Eric. "SolarCity Acquires Zep Solar for $158 Million." *Greentech Media*. Greentech Media, 09 Oct. 2013. Web. <https://www

.greentechmedia.com/articles/read/SolarCity-Acquires-Zep
-Solar-For-158-Million>.

26. *Global Off-Grid Solar Market Report*. Rep. GOGLA, Dec. 2020. Web. <https://www.gogla.org/sites/default/files/resource_docs/global_off-grid_solar_market_report_h2_2020.pdf>.

27. "All You Need to Know About Home Geothermal Heating & Cooling." *Dandelion Energy*. 13 Mar. 2020. Web. <https://dandelionenergy.com/all-you-need-to-know-about-home-geothermal-heating-cooling>.

28. "Original Company." *Jacobs Wind Electric*. 2019. Web. <http://www.jacobswind.net/history/original-company>.

29. "Rivers & Streams." EPA, 13 Mar. 2013. Web. <https://archive.epa.gov/water/archive/web/html/index-17.html>.

30. "United States Surpasses 2 Million Solar Installations." *SEIA*. Solar Energy Industries Association, 09 May 2019. Web. <www.seia.org/news/united-states-surpasses-2-million-solar-installations>.

31. 5 GW of residential and C&I solar were added in 2020 and another 5 to 7 GW is estimated for 2021. Assuming an average 8 kW per rooftop installation, this adds 1,375,000 new installations in 2020 and 2021.
    Solar Market Insight Report 2020 Year in Review. Rep. SEIA, 16 Mar. 2021. Web. <https://www.seia.org/research-resources/solar-market-insight-report-2020-year-review>.

32. Kennedy, Brian, and Cary Lynne Thigpen. "More U.S. Homeowners Say They Are Considering Home Solar Panels." *Pew Research Center*, Pew Research Center, 14 Aug. 2020. Web. <www.pewresearch.org/fact-tank/2019/12/17/more-u-s-homeowners-say-they-are-considering-home-solar-panels/>.

33. Heeter, Jenny, and Gabriel Chan. "Sharing the Sun Community Solar Project Data (June 2020)." National Renewable Energy Laboratory, June 2020. Web. <https://data.nrel.gov/submissions/149>.

34. Lambert, Fred. "Tesla Has Doubled Number of Powerwalls Installed to 200,000 over the Last Year." *Electrek*. 26 May 2021. Web. <https://electrek.co/2021/05/26/tesla-doubled-powerwalls-installed-over-last-year>.

CHAPTER 4
FROM FUELS TO TECHNOLOGIES

1. "The Vagabonds." *The Henry Ford.* Web. <https://www.thehenry-ford.org/collections-and-research/digital-resources/popular-topics/the-vagabonds/>.

2. Newton, James D. Uncommon Friends: Life with Thomas Edison, Henry Ford, Harvey Firestone, Alexis Carrel, and Charles Lindbergh. San Diego: Harcourt Brace Jovanovich, 1989. Print. This citation is lightly edited for brevity.

3. "(23000 TW years in exajoules)/12." *WolframAlpha Computational Intelligence.* Web. <https://www.wolframalpha.com/input/?i=%282 3000+tw+years++in+exajoules%29%2F12>.

4. "Total Solar Energy Received by the Earth Is 201,480,000,000,000,000 KWh/ Year." *Freeing Energy.* 2021. Web. <https://www.freeingenergy.com/m181/>.

5. "Annual Energy Output of a 1 GW Nuclear Plant." *Freeing Energy.* Web. <https://www.freeingenergy.com/m193/>.

6. "The Annual Global Electricity Consumption Is 167,718 terawatt hours." *Freeing Energy.* Web. <https://www.freeingenergy.com/m180/>.

7. Perez, Marc, and Richard Perez. "Update 2015—A Fundamental Look at Supply Side Energy Reserves for the Planet." *IEA SHCP Newsletter* 62 (Nov. 2015). Web. <http://research.asrc.albany.edu/people/faculty/perez/2015/IEA.pdf>.

8. "What Is U.S. Electricity Generation by Energy Source?" *Frequently Asked Questions (FAQS).* EIA. Web. <https://www.eia.gov/tools/faqs/faq.php?id=427&t=3>.

9. "List of Solar Thermal Power Stations." *Wikipedia.* Wikimedia Foundation. Web. <https://en.wikipedia.org/wiki/List_of_solar_thermal_power_stations>.

10. "Renewable Energy Cost of Electricity Is Far Cheaper than Fossil Fuels." *Freeing Energy.* Web. <https://www.freeingenergy.com/g108/>.

11. *Levelized Cost of Energy Analysis – Version 4.0.* Rep. Lazard, June 2010. Web. <https://www.puc.pa.gov/pcdocs/1173215.pdf>.

12. "Albert Einstein." *Wikipedia.* Wikimedia Foundation. Web. <https://en.wikipedia.org/wiki/Albert_Einstein#1905_ –_Annus_Mirabilis_papers>.

13. Perlin, John. *Let It Shine: The 6,000-Year Story of Solar Energy.* Novato, CA: New World Library, 2013. 478. Print.

14. Perlin, John. *Let It Shine: The 6,000-Year Story of Solar Energy.* Novato, CA: New World Library, 2013. 317. Print.

15. "Sputnik 1." *Wikipedia.* Wikimedia Foundation. Web. <https://en.wikipedia.org/wiki/Sputnik_1>.

16. "Explorer 1." *Wikipedia.* Wikimedia Foundation. Web. <https://en.wikipedia.org/wiki/Explorer_1>.

17. Hsu, Andrea. "How Big Oil Of The Past Helped Launch The Solar Industry Of Today." *NPR.* 30 Sept. 2019. Web. <https://www.npr.org/2019/09/30/763844598/how-big-oil-of-the -past-helped-launch-the-solar-industry-of-today>.

18. "UNIVAC I." *Wikipedia.* Wikimedia Foundation. Web. <https://en.wikipedia.org/wiki/UNIVAC_I>.

19. Hinum, Klaus. "Apple M1 Processor—Benchmarks and Specs." *Notebookcheck.* 10 Nov. 2020. Web. <https://www.notebookcheck. net/Apple-M1-Processor-Benchmarks-and-Specs.503613.0.html>.

20. Moore, Gordon E. "Cramming More Components onto Integrated Circuits." Reprinted from Electronics, Volume 38, Number 8, April 19, 1965, Pp.114 Ff." *IEEE Solid-State Circuits Society Newsletter* 11.3 (2006): 33-35. Web. <https://ieeexplore.ieee.org/ document/4785860>.

21. "Global Power Plant Database." *World Resources Institute.* 02 June 2021. Web. <https://datasets.wri.org/dataset/globalpowerplantdatabase>. "Global Coal Plant Tracker." *End Coal.* Web. <https://endcoal.org/ global-coal-plant-tracker/>. *World Nuclear Performance Report 2020.* Rep. World Nuclear Association, 2020. Web. <https://www.world-nuclear.org/

our-association/publications/global-trends-reports/world-nuclear-performance-report.aspx>.

22. There were 341,000 wind turbines at the end of 2016; the 2016 capacity was 488,508 MW. The 2019 capacity was 650,758 MW. Therefore, we can extrapolate the wind turbine count to be 341,000*(650,758/448,508)=454,257, or approximately 500,000 wind turbines.
Frangoul, Anmar. "There Are over 341,000 Wind Turbines on the Planet: Here's How Much of a Difference They're Actually Making." *CNBC*. 08 Sept. 2017. Web. <https://www.cnbc.com/2017/09/08/there-are-over-341000-wind-turbines-on-the-planet-why-they-matter.html>.

23. "Snapshot 2020—IEA-PVPS." *IEA*. IEA Photovoltaic Power Systems Programme, 29 Apr. 2020. Web. <https://iea-pvps.org/snapshot-reports/snapshot-2020/>.

24. "Goals of the Solar Energy Technologies Office." *Energy.gov*. Solar Energy Technologies Office. Web. <https://www.energy.gov/eere/solar/goals-solar-energy-technologies-office>.

25. "Energy Storage Grand Challenge Roadmap." *Energy.gov*. 21 Dec. 2020. Web. <https://www.energy.gov/energy-storage-grand-challenge/articles/energy-storage-grand-challenge-roadmap>.

26. Note: There is one exception. Nuclear breeder reactors allow for spent fuel to be used a second time. These types of plants have many issues; only France operates them today.

27. Note: While there are many small solar projects, most of the solar power generated in the US still comes from large, utility-scale projects (at least for now). For these larger projects, a 2019 Berkeley Labs report found that 70% use moving parts called trackers, which move solar panels across the day to optimize their position toward the sun.

28. *Annual Energy Outlook 2005*. Rep. DOE/EIA, Feb. 2005. Web. Table A16. <https://www.eia.gov/outlooks/archive/aeo05/pdf/0383(2005).pdf>.

29. "Solar Industry Research Data." *SEIA*. Web. <https://www.seia.org/solar-industry-research-data>. Retrieved July 16, 2021.

30. Shahan, Zachary. "IEA Gets Hilariously Slammed For Obsessively Inaccurate Renewable Energy Forecasts." *CleanTechnica*. 06 Sept. 2017. Web. <https://cleantechnica.com/2017/09/06/iea-gets-hilariously-slammed-continuously-pessimistic-renewable-energy-forecasts/>.

31. Wind and Solar Data and Projections from the U.S. Energy Information Administration: Past Performance and Ongoing Enhancements. Rep. EIA, Mar. 2016. Web. <https://www.eia.gov/outlooks/aeo/supplement/renewable/pdf/projections.pdf>.

32. "Solar Chernobyl Project." *Solar Chernobyl*. Web. <https://solarchernobyl.com/>.

33. Sunder, Kalpana. "The 'Solar Canals' Making Smart Use of India's Space." *BBC Future*. BBC, 3 Aug. 2020. Web. <https://www.bbc.com/future/article/20200803-the-solar-canals-revolutionising-indias-renewable-energy>.

34. Gagnon, Pieter, Robert Margolis, Jennifer Melius, Caleb Phillips, and Ryan Elmore. *Rooftop Solar Photovoltaic Technical Potential in the United States: A Detailed Assessment*. Rep. NREL, Jan. 2016. Web. <www.nrel.gov/docs/fy16osti/65298.pdf>.

35. Spencer, Robert S., Jordan Macknick, Alexandra Aznar, Adam Warren, and Matthew O. Reese. "Floating Photovoltaic Systems: Assessing the Technical Potential of Photovoltaic Systems on Man-Made Water Bodies in the Continental United States." *Environmental Science & Technology* 53.3 (2018): 1680-689. Web. < https://www.osti.gov/pages/biblio/1489330>.

36. Kimmelman, Michael. "Paved, but Still Alive." *The New York Times*. 06 Jan. 2012. Web. <https://www.nytimes.com/2012/01/08/arts/design/taking-parking-lots-seriously-as-public-spaces.html>.

37. "This Month in Physics History." *American Physical Society*. Web. <https://www.aps.org/publications/apsnews/200603/history.cfm>.

38. "Scientists Win Nobel Prize in Chemistry for Work on Lithium-ion Battery." *The Washington Post*. 09 Oct. 2019. Web. <https://www.washingtonpost.com/lifestyle/kidspost/

scientists-win-nobel-prize-in-chemistry-for-work-on-lithium-ion-batt
ery/2019/10/09/9c2e3ebe-e481-11e9-a331-2df12d56a80b_story.html>.

39. LeVine, Steve. "The man who brought us the lithium-ion battery
     at the age of 57 has an idea for a new one at 92." *Quartz*. 5 Feb.
     2015. Web. <https://qz.com/338767/the-man-who-brought-us-the-
     lithium-ion-battery-at-57-has-an-idea-for-a-new-one-at-92/>.

40. "The History of the Electric Car." *Energy.gov*. DOE, 15 Sept. 2014.
     Web. <https://www.energy.gov/articles/history-electric-car>.

41. Gilmore, C. P. "Electric Autos . . . They're on the Way." *Popular
     Science*. Dec. 1966. Web. <https://books.google.com/books?id=lyk
     DAAAAMBAJ&pg=PA79&lpg=PA79&dq=popular+science+ford+ba
     ttery+1966+sulfur#v=onepage&q=popular%20science%20ford%20
     battery%201966%20sulfur&f=false>.

42. "General Motors EV1." *Wikipedia*. Wikimedia Foundation. Web.
     <https://en.wikipedia.org/wiki/General_Motors_EV1>.

43. "Tesla Battery Day." Tesla, 22 Sept. 2020. Web. <https://tesla-cdn.
     thron.com/static/VM2Z8A_BD_Keynote_final_PSPS0F.pdf>.

44. "The US Uses an Average of 3,600 Gigawatt Hours of Electricity
     Each Night." *Freeing Energy*. Web. <https://www.freeingenergy.
     com/m182/>.

45. Baker, David R. "Electric Cars Closing In on Gas Guzzlers as
     Battery Costs Plunge." *Bloomberg Green*. Bloomberg, 16 Dec. 2020.
     Web. <https://www.bloomberg.com/news/articles/2020-12-16/
     electric-cars-closing-in-on-gas-guzzlers-as-battery-costs-plunge>.

46. Ziegler, Micah S., Joshua M. Mueller, Gonçalo D. Pereira, Juhyun
     Song, Marco Ferrara, Yet-Ming Chiang, and Jessika E. Trancik.
     "Storage Requirements and Costs of Shaping Renewable Energy
     Toward Grid Decarbonization." *Joule* 3.9 (2019): 2134-153. 18 Sept.
     2019. Web. <https://www.sciencedirect.com/science/article/pii/
     S2542435119303009>.

47. Potter, Ben. "Tesla Battery Boss: We Can Solve SA's Power
     Woes in 100 Days." *Australian Financial Review*. 09 Mar.
     2017. Web. <https://www.afr.com/politics/tesla-battery
     -boss-we-can-solve-sas-power-woes-in-100-days-20170308-gut8xh>.

48. "Hornsdale Power Reserve." *Wikipedia*. Wikimedia Foundation. Web. <https://en.wikipedia.org/wiki/Hornsdale_Power_Reserve>.

49. Fung, Brian. "Tesla's Enormous Battery in Australia, Just Weeks Old, Is Already Responding to Outages in 'record' Time." *The Washington Post*. 26 Dec. 2017. Web. <https://www.washingtonpost.com/news/the-switch/wp/2017/12/26/teslas-enormous-battery-in-australia-just-weeks-old-is-already-responding-to-outages-in-record-time/>.

50. Parkinson, Giles. "Revealed: True Cost of Tesla Big Battery, and Its Government Contract." *RenewEconomy*. 21 Sept. 2018. Web. <https://reneweconomy.com.au/revealed-true-cost-of-tesla-big-battery-and-its-government-contract-66888/>.

51. Maisch, Marija. "South Australia's Tesla Big Battery Saves $40 Million in Grid Stabilization Costs." *PV Magazine International*. 05 Dec. 2018. Web. <https://www.pv-magazine.com/2018/12/05/south-australias-tesla-big-battery-saves-40-million-in-grid-stabilization-costs/>.

52. "US Per Capita Electricity Consumption Is Declining." *Freeing Energy*. Web. <https://www.freeingenergy.com/g101/>.

53. *2019 Lighting R&D Opportunities*. Rep. DOE, Jan. 2020. Web. <https://www.energy.gov/sites/prod/files/2020/01/f70/ssl-rd-opportunities2-jan2020.pdf>.

54. *Lighting—Analysis*. Rep. IEA, June 2020. Web. <https://www.iea.org/reports/lighting>.

CHAPTER 5
HIDDEN PATTERNS OF INNOVATION

1. Gaddy, Benjamin, Varun Sivaram, Francis O'Sullivan. "Venture Capital and Cleantech: The Wrong Model for Clean Energy Innovation." *MIT Energy Initiative*. Jul. 2016. <https://energy.mit.edu/wp-content/uploads/2016/07/MITEI-WP-2016-06.pdf>.

2. "Solyndra—Funding, Financials, Valuation & Investors." *Crunchbase*. Web. <https://www.crunchbase.com/organization/solyndra/company_financials>.

3.   Deign, Jason. "What Happened to the VC-Backed Startup Class of 2010?" *Greentech Media*. 10 Nov. 2020. Web. <https://www.greentechmedia.com/articles/read/whatever-happened-to-the-vc-backed-startup-class-of-2010>.

4.   "Prices Flat in Polysilicon Market." *The Washington Post*. 23 July 2013. Web. <https://www.washingtonpost.com/business/economy/prices-flat-in-polysilicon-market/2013/07/23/914479d0-f3e4-11e2-9434-60440856fadf_graphic.html>.

5.   "Latest Solar PV News." *PVinsights*. Web. <http://pvinsights.com/>. Retrieved May 5, 2021.

6.   "Henry Hub Natural Gas Spot Price (Dollars per Million Btu)." *Independent Statistics & Analysis*. EIA. Web. <https://www.eia.gov/dnav/ng/hist/rngwhhdM.htm>.

7.   "Perovskite Solar Cell." *Wikipedia*. Wikimedia Foundation. Web. <https://en.wikipedia.org/wiki/Perovskite_solar_cell>.

8.   Nussey, Bill. "Jeff Chamberlain: Are batteries the final puzzle piece for a clean energy grid?" Freeing Energy Podcast, 08 May 2019. < https://www.freeingenergy.com/podcast-011-jeff-chamberlain-are-batteries-the-final-puzzle-piece-for-a-clean-energy-grid/>.

9.   "Battery Pack Prices Cited Below $100/kWh for the First Time in 2020, While Market Average Sits at $137/kWh." *BloombergNEF*. 16 Dec. 2020. Web. <https://about.bnef.com/blog/battery-pack-prices-cited-below-100-kwh-for-the-first-time-in-2020-while-market-average-sits-at-137-kwh/>.

10.   Temple, James. "Inside the Fall, and Rebirth, of a Bill Gates–Backed Battery Startup." *MIT Technology Review*. 08 Aug. 2017. Web. <https://www.technologyreview.com/2017/08/08/150080/inside-the-fall-and-rebirth-of-a-bill-gates-backed-battery-startup/>.

11.   Szmigiera, M. "Manufacturing Labor Costs per Hour: China, Vietnam, Mexico 2016-2020." *Statista*. 30 Mar. 2021. Web. <https://www.statista.com/statistics/744071/manufacturing-labor-costs-per-hour-china-vietnam-mexico/>.

12.   Woodhouse, Michael, Brittany Smith, Ashwin Ramdas, and Robert Margolis. Crystalline Silicon Photovoltaic Module Manufacturing

Costs and Sustainable Pricing: 1H 2018 Benchmark and Cost Reduction Road Map. Rep. NREL, 2019. Web. <https://www.nrel.gov/docs/fy19osti/72134.pdf>.

13. Kennedy, Danny. Personal interview. 03 Jan. 2018.

14. 6000 firms employing 124,594 people is 20.7 per firm; *Solar Job Census*. Rep. IREC, 11 Jan. 2021. Web. July 2021. <https://irecusa.org/programs/solar-jobs-census/>; "Total Number of Installation firms for Utility, Commercial, and Residential Solar." *Freeing Energy*. Web. <https://www.freeingenergy.com/m190/>.

15. "Greener Office Buildings with Power from Façade and Rooftop Photovoltaic Systems." *ZSW*. 24 Feb. 2021. Web. <https://www.zsw-bw.de/en/newsroom/news/news-detail/news/detail/News/greener-office-buildings-with-power-from-facade-and-rooftop-photovoltaic-systems.html>.

16. "Most Pumped Storage Electricity Generators in the U.S. Were Built in the 1970s." Today in Energy. *Independent Statistics & Analysis*. EIA. Web. <https://www.eia.gov/todayinenergy/detail.php?id=41833>.

17. Lambert, Fred. "Tesla Has Doubled Number of Powerwalls Installed to 200,000 over the Last Year." *Electrek*. 26 May 2021. Web. <https://electrek.co/2021/05/26/tesla-doubled-powerwalls-installed-over-last-year>.

18. Evans, Scott. "2013 Tesla Model S Beats Chevy, Toyota, and Cadillac for Ultimate Car of the Year Honors." *MotorTrend*. 10 July 2019. Web. <https://www.motortrend.com/news/2013-tesla-model-s-beats-chevy-toyota-cadillac-ultimate-car-of-the-year/>.

CHAPTER 6
BILLION DOLLAR DISRUPTIONS

1. "Wright Brothers." *Wikipedia*. Wikimedia Foundation. Web. <https://en.wikipedia.org/wiki/Wright_brothers>.

2. Newcomb, Simon. *Side-lights on Astronomy and Kindred Fields of Popular Science*. 2012. Project Gutenberg, 13 June 2009. Web. <https://www.gutenberg.org/files/4065/4065-h/4065-h.htm>.

3. "The War of the Currents: AC vs. DC Power." *Energy.gov.* DOE, 18 Nov. 2014. Web. <https://www.energy.gov/articles/war-currents -ac-vs-dc-power>.

4. "US Per Capita Electricity Consumption Is Declining." *Freeing Energy.* Web. <https://www.freeingenergy.com/g101/>.

5. Hoskins, Anne. "What Value Do Batteries Bring to California Homeowners?" *Solar Power World.* 13 Jan. 2020. Web. <https:// www.solarpowerworldonline.com/2020/01/what-value-do -batteries-bring-to-california-homeowners/>.

6. Fox-Penner, Peter S., James E. Rogers, Daniel C. Esty, Daniel Dobbeni, and Lyndon Rive. *Smart Power: Climate Change, the Smart Grid, and the Future of Electric Utilities.* Washington: Island, 2014. Print.

7. "Aggregated Reports." *Aggregated Reports—TLC.* NYC Taxi & Limousine Commission. Web. <https://www1.nyc.gov/site/tlc/ about/aggregated-reports.page>.

8. "Natural Gas Annual 2019 (NGA)—Energy Information Administration—With Data for 2019." *Independent Statistics & Analysis.* EIA. Web. <https://www.eia.gov/naturalgas/annual/>.

9. "Gasoline Station Sales in the United States from 1992 to 2019." *Statista.* Web. <https://www.statista.com/statistics/197637/annual- gasoline-station-sales-in-the-us-since-1992/>. Retrieved January 2021.

10. "Electric Power Monthly." EIA, Feb. 2021. Web. <https://www.eia. gov/electricity/monthly/epm_table_grapher.php?t=epmt_5_02>. Retrieved February 2021.

11. *Global Hydrogen Generation Market Size Report, 2021-2028.* Rep. Grand View Research, Mar. 2021. Web. <https://www.grandviewre- search.com/industry-analysis/hydrogen-generation-market>.

12. "Energy Density." *Wikipedia.* Wikimedia Foundation. Web. <https:// en.wikipedia.org/wiki/Energy_density>.

13. "Hydrogen Safety." *Wikipedia.* Wikimedia Foundation. Web. <https:// en.wikipedia.org/wiki/Hydrogen_safety>.

14. Ramkumar, Amrith. "Concentrated Solar Power Firm Heliogen to Go Public in $2 Billion SPAC Merger." *The Wall Street Journal.* 06 July

2021. Web. <https://www.wsj.com/articles/concentrated-solar-power-firm-heliogen-to-go-public-in-2-billion-spac-merger-11625607000>.

15. Harder, Amy. "In Pandemic's Wake, Global Support Builds for Hydrogen." *Axios.* 27 July 2020. Web. <https://www.axios.com/hydrogen-support-growing-5ee4cfbb-ee83-4494-8561-c172467b7e5e.html>.

16. Blank, Thomas Koch. "Technology Disruption in the Global Steel Industry." *RMI.* 02 Dec. 2020. Web. <https://rmi.org/technology-disruption-in-the-global-steel-industry/>.

17. Connelly, Elizabeth, Amgad Elgowainy, and Mark Ruth. *DOE Hydrogen and Fuel Cells Program Record.* Rep. DOE, 01 Oct. 2019. Web. <https://www.hydrogen.energy.gov/pdfs/19002-hydrogen-market-domestic-global.pdf>.

18. *The Future of Hydrogen.* Rep. IEA, June 2019. Web. <https://www.iea.org/reports/the-future-of-hydrogen>.

19. "Global Energy Perspective 2019: Reference Case." McKinsey, Jan. 2019. Web. <https://www.mckinsey.com/~/media/McKinsey/Industries/Oil%20and%20Gas/Our%20Insights/Global%20Energy%20Perspective%202019/McKinsey-Energy-Insights-Global-Energy-Perspective-2019_Reference-Case-Summary.ashx>.

20. *Electric Vehicle Outlook 2021.* Rep. BloombergNEF. Web. <https://about.bnef.com/electric-vehicle-outlook/>.

21. "How Much Time Do Americans Spend Behind the Wheel?" *Volpe National Transportation Systems Center.* U.S. Department of Transportation, 11 Dec. 2017. Web. <https://www.volpe.dot.gov/news/how-much-time-do-americans-spend-behind-wheel>.

22. Hanley, Steve. "All Volkswagen MEB-Based Electric Cars Will Be V2G Capable Beginning In 2022." *CleanTechnica.* 06 Apr. 2021. Web. <https://cleantechnica.com/2021/04/06/all-volkswagen-meb-based-electric-cars-will-be-v2g-capable-beginning-in-2022/>.

23. LeVine, Steve. "The Million-mile Battery Is Really HERE. It Will Change Everything." *The Electric.* The Information. Web. <https://www.theinformation.com/newsletters/the-electric/archive/65e4692f-6a19-441e-aa20-d1bf96d0e450>.

24. Johnson, Lacey. "EVs Could Save California Billions in Energy Storage Investment." *Greentech Media.* 29 May 2018. Web. <https://www.greentechmedia.com/articles/read/ev-save-california-billions-energy-storage-investment#gs.y33xm0>.

25. Gross, Andrew. "New Study Reveals When, Where and How Much Motorists Drive." *AAA Newsroom.* 16 Apr. 2015. Web. <https://newsroom.aaa.com/2015/04/new-study-reveals-much-motorists-drive/>.

26. Lips, Brian, Autumn Proudlove, and David Sarkisian. *50 States of Electric Vehicles.* Rep. NC Clean Energy Technology Center, Aug. 2020. Web. <https://nccleantech.ncsu.edu/wp-content/uploads/2020/08/Q2-20_EV_execsummary_Final.pdf>.

27. Blumberg, Stephen J., and Julian V. Luke. *Wireless Substitution: Early Release of Estimates From the National Health Interview Survey, January–June 2018.* Rep. National Center for Health Statistics, Dec. 2018. Web. <https://www.cdc.gov/nchs/data/nhis/earlyrelease/wireless201812.pdf>.

28. Nussey, Bill. "A Simple and Affordable Plan to Address the Intermittency of Solar and Wind." *Freeing Energy.* 23 July 2019. Web. <https://www.freeingenergy.com/solar-wind-intermittency-batteries-curtail-overbuild/>.

29. "Solar Data Cheat Sheet." *SEIA.* 16 Mar. 2021. Web. 22 July 2021. <https://www.seia.org/research-resources/solar-data-cheat-sheet>.

30. Wiser, Ryan, Mark Bolinger, and Joachim Seel. Benchmarking Utility-Scale PV Operational Expenses and Project Lifetimes: Results from a Survey of U.S. Solar Industry Professionals. Tech. Lawrence Berkeley National Laboratory, June 2020. Web. <https://eta-publications.lbl.gov/sites/default/files/solar_life_and_opex_report.pdf>.

31. "Total US Installed Solar for 2020 Could Power 19 Million Homes." *Freeing Energy.* Web. <https://www.freeingenergy.com/m195/>.

32. Masanet, Eric, Arman Shehabi, Nuoa Lei, Sarah Smith, and Jonathan Koomey. "Recalibrating Global Data Center Energy-use Estimates." *Science.* American Association for the Advancement

of Science, 28 Feb. 2020. Web. <https://science.sciencemag.org/content/367/6481/984>.

33. Service, Robert F. "Ammonia-A Renewable Fuel Made from Sun, Air, and Water-Could Power the Globe Without Carbon." *Science.* 12 July 2018. Web. <https://www.sciencemag.org/news/2018/07/ammonia-renewable-fuel-made-sun-air-and-water-could-power-globe-without-carbon>.

34. *Energy Access Outlook 2017.* Rep. IEA, Oct. 2017. Web. <https://www.iea.org/reports/energy-access-outlook-2017>.

35. Riggs, John A. High Tension: FDR's Battle to Power America. Diversion, 2020. Print.

36. George, Libby. "Nigeria's Diesel-dependent Economy Braces for Clean-fuel Rules." *Reuters.* 18 Sept. 2019. Web. <https://www.reuters.com/article/us-nigeria-power-diesel/nigerias-diesel-dependent-economy-braces-for-clean-fuel-rules-idUSKBN1W323K>.

37. Olurounbi, Ruth, and William Clowes. "Nigeria Fuel Subsidies Near $300 Million a Month, NNPC Says." *Bloomberg.* 26 Mar. 2021. Web. <https://www.bloomberg.com/news/articles/2021-03-25/nigeria-fuel-subsidy-hits-nearly-300-million-a-month-nnpc-says>.

38. "BBOXX—Crunchbase Company Profile & Funding." *Crunchbase.* Web. <https://www.crunchbase.com/organization/bboxx>.

39. *Power Africa in Rwanda.* Rep. USAID, Sept. 2016. Web. <https://www.usaid.gov/sites/default/files/documents/1860/RwandaCountryFactSheet__2016.09%20FINAL.pdf>.

40. "Off-grid Electrification Helping to Achieve Rwanda's Energy Targets." *Rwanda Energy Group.* 06 May 2020. Web. <https://www.reg.rw/media-center/news-details/news/off-grid-electrification-helping-to-achieve-rwandas-energy-targets/>.

41. *Global Off-Grid Solar Market Report Semi-Annual Sales and Impact Data.* Rep. GOGLA, June 2020. Web. <https://www.gogla.org/sites/default/files/resource_docs/global_off_grid_solar_market_report_h1_2020.pdf>.

CHAPTER 7
UTILITIES VS THE FUTURE

1. Wesoff, Eric. "Breaking: FERC Dismisses Petition by Shadowy Group to End Net Metering." *PV Magazine USA*. 16 July 2020. Web. <https://pv-magazine-usa.com/2020/07/16/ferc-dismisses-petition-by-shadowy-group-to-end-net-metering/>.

2. Sundback, Mark, Bill Rappolt, and Andrew Mina. "FERC Rejects Net Metering Petition, But Fight Is Far From Over." *Energy Law Blog*. 17 July 2020. Web. <https://www.energylawinfo.com/2020/07/ferc-rejects-net-metering-petition-nera/>.

3. Olasky, Marvin N. "Hornswoggled! How Ma Bell and Chicago Ed Conned Our Grandparents and Stuck Us with the Bill." *Reason*. Feb. 1986. Web. <https://reason.com/1986/02/01/hornswoggled/>.

4. "Special Interests Behind Anti-Solar Ballot Initiative Consumers for Smart Solar." *Energy and Policy Institute*. 16 Nov. 2015. Web. <https://www.energyandpolicy.org/special-interests-behind-anti-solar-ballot-initiative-consumers-for-smart-solar/>.

5. Klas, Mary Ellen. "Insider Reveals Deceptive Strategy behind Florida's Solar Amendment." *Miami Herald*. 18 Oct. 2016. Web. <https://www.miamiherald.com/news/politics-government/election/article109017387.html>.

6. "Consumers for Smart Solar." *Ballotpedia*. Web. <https://ballotpedia.org/Consumers_for_Smart_Solar>.

7. "Interconnection Guidelines." *DSIRE*. 24 July 2020. Web. <https://programs.dsireusa.org/system/program/detail/782>.

8. "Alabama Power Customers Question 'Punitive' Rooftop Solar Charges." *Energy News Network*. Energywire, 25 Nov. 2019. Web. <https://energynews.us/2019/11/25/alabama-power-customers-question-punitive-rooftop-solar-charges/>.

9. Baldwin, John. "How Going Solar Could Impact Your Homeowner's Insurance." *Solar United Neighbors*. 18 Jan. 2019. Web. <https://www.solarunitedneighbors.org/news/how-going-solar-could-impact-your-homeowners-insurance/>.

10. Opalka, Bill. "Solar Advocates Look to Repeal Maine's 'fundamentally Unjust' Solar Rules." *Energy News Network*. 19 Feb. 2019. Web. <https://energynews.us/2019/02/19/solar-advocates-look-to-repeal-maines-fundamentally-unjust-solar-rules/>.

11. Branscomb, Lewis M., and Philip E. Auerswald. Taking Technical Risks: How Innovators, Managers, and Investors Manage Risk in High-Tech Innovations. Cambridge, MA, USA: MIT, 2003. Print.

12. Costello, Ken. *Research and Development by Public Utilities: Should More Be Done?* National Regulatory Research Institute, 08 Nov. 2015. Web. <https://pubs.naruc.org/pub.cfm?id=4AA29DB3-2354-D714-51DB-4CFE5EFE50A7>.
    "Domestic R&D Paid for by the Company and Others and Performed by the Company as a Percentage of Domestic Net Sales, by Industry and Company Size: 2017." *National Center for Science and Engineering Statistics.* NSF, 2017. Web. <https://ncses.nsf.gov/pubs/nsf20311/assets/data-tables/tables/nsf20311-tab017.pdf>.
    This is a contentious point. Therefore, two sources have been included.

13. Fox-Penner, Peter S. *Power after Carbon: Building a Clean, Resilient Grid.* Cambridge, MA: Harvard UP, 2020. Print.

14. "Industries." *OpenSecrets.* Center for Responsive Politics, 30 Apr. 2021. Web. <http://www.opensecrets.org/federal-lobbying/industries?cycle=a>.

15. Kind, Peter. "Disruptive Challenges: Financial Implications and Strategic Responses to a Changing Retail Electric Business." Edison Electric Institute, Jan. 2013. Web. <https://web.archive.org/web/20170829023930/http://www.eei.org/ourissues/finance/Documents/disruptivechallenges.pdf>.

16. "2017-10-30 Comment Response to the Published Notice of Request for Information (RFI)." *Regulations.gov.* Energy Efficiency and Renewable Energy Office, 31 Oct. 2017. Web. <https://www.regulations.gov/comment/EERE-2017-OT-0056-0059>.

17. Hayibo, Koami Soulemane, and Joshua M. Pearce. "A Review of the Value of Solar Methodology with a Case Study of the U.S. VOS."

Renewable and Sustainable Energy Reviews 137 (2021): 110599. Web. <https://www.sciencedirect.com/science/article/abs/pii/S1364032120308832>.

18. "Form EIA-861M (formerly EIA-826) Detailed Data." *Independent Statistics & Analysis*. EIA, 29 Oct. 2020. Web. <https://www.eia.gov/electricity/data/eia861m/>.

19. "Reduction in US electric utilities' revenue from distributed rooftop solar." *Freeing Energy*. Web. <https://www.freeingenergy.com/m121/>.
    "Electric Sales, Revenue, and Average Price." *Independent Statistics & Analysis*. EIA, 06 Oct. 2020. Web. <https://www.eia.gov/electricity/sales_revenue_price/>.

20. *US Light Bulb Standards Save Billions for Consumers But Manufacturers Seek a Rollback*. Issue brief. ASAP/ACEEE, July 2018. Web. <https://appliance-standards.org/sites/default/files/light_bulb_brief_2.pdf>.

21. Barbose, Galen L. *Putting the Potential Rate Impacts of Distributed Solar into Context*. Rep. no. LBNL-1007060. Lawrence Berkeley National Laboratory, Jan. 2017. Web. <https://emp.lbl.gov/publications/putting-potential-rate-impacts>.

22. "Net Metering." *SEIA*. Web. <https://www.seia.org/initiatives/net-metering>.

23. "Programs." *DSIRE*. NC Clean Energy Technology Center. Web. <https://programs.dsireusa.org/system/program?type=37&>.

24. Shea, Ryan, Arthur Coulston, and Rushad Nanavatty. "How COVID-19 Is Pushing Cities to Change Solar Permitting for the Better." *RMI*. 16 July 2020. Web. <https://rmi.org/how-covid-19-is-pushing-cities-to-change-solar-permitting-for-the-better/>.

25. Birch, Andrew. "How to Halve the Cost of Residential Solar in the US." *Greentech Media*. 05 Jan. 2018. Web. <https://www.greentechmedia.com/articles/read/how-to-halve-the-cost-of-residential-solar-in-the-us>.

26. "U.S. Department of Energy SunShot Initiative." *Energy.gov*. DOE. Web. <https://www.energy.gov/sites/prod/files/2016/05/f32/SC%20Fact%20Sheet-508.pdf>.

27. "Over-the-Air Reception Devices Rule." *Federal Communications Commission*. Web. <https://www.fcc.gov/media/over-air-reception-devices-rule>.

28. "Time of Use and Dynamic Pricing Rates in the US." *European University Institute*. Web. <https://fsr.eui.eu/time-of-use-and-dynamic-pricing-rates-in-the-us/>.

29. "2019 Building Energy Efficiency Standards." California Energy Commission, 01 Mar. 2021. Web. <https://web.archive.org/web/20210301191939/https://ww2.energy.ca.gov/title24/2019standards/documents/Title24_2019_Standards_detailed_faq.pdf>.

30. Molle, Grégoire. "How Blockchain Helps Brooklyn Dwellers Use Neighbors' Solar Energy." *NPR*. 04 July 2016. Web. <https://www.npr.org/sections/alltechconsidered/2016/07/04/482958497/how-blockchain-helps-brooklyn-dwellers-use-neighbors-solar-energy>.

31. "Brooklyn Microgrid Selected to Compete in $5m U.S. Energy Dept Program to Open up New Opportunities for Solar Power." *LO3 Energy*. 31 May 2017. Web. <https://lo3energy.com/brooklyn-microgrid-selected-compete-5m-u-s-energy-dept-program-open-new-opportunities-solar-power/>.

32. Maloney, Peter. " Brooklyn Microgrid Launches Campaign to Create Regulatory Sandbox." *Microgrid Knowledge*. 18 Oct. 2019. Web. <https://microgridknowledge.com/brooklyn-microgrid-regulatory-sandbox/>.

33. "California Crisis Timeline." *Frontline*. Public Broadcasting Service. Web. <https://www.pbs.org/wgbh/pages/frontline/shows/blackout/california/timeline.html>.

34. "Investor-owned Utilities Served 72% of U.S. Electricity Customers in 2017." Today in Energy. *Independent Statistics & Analysis*. EIA, 15 Aug. 2019. Web. <https://www.eia.gov/todayinenergy/detail.php?id=40913>.

35. "Public Service Commissioner (state Executive Office)." *Ballotpedia*. Web. <https://ballotpedia.org/Public_Service_Commissioner_(state_executive_office)>.

36. "Electric Power Monthly." EIA. Web. <https://www.eia.gov/electricity/monthly/epm_table_grapher.php?t=epmt_5_01>.

37. Parker, Kim, Juliana Menasce Horowitz, Anna Brown, Richard Fry, D'Vera Cohn, and Ruth Igielnik. "Demographic and Economic Trends in Urban, Suburban and Rural Communities." *Pew Research Center's Social & Demographic Trends Project*. Pew Research Center, 22 May 2018. Web. <https://www.pewresearch.org/social-trends/2018/05/22/demographic-and-economic-trends-in-urban-suburban-and-rural-communities/>.

38. Mulkern, Anne C. "A Solar Boom So Successful, It's Been Halted." *Scientific American*. 20 Dec. 2013. Web. <https://www.scientificamerican.com/article/a-solar-boom-so-successfull-its-been-halted/>.

39. Storrow, Benjamin. "Energy Transitions: More Coal Has Retired under Trump than in Obama's 2nd Term." *E&E News*. 22 June 2020. Web. <https://www.eenews.net/stories/1063430425>.

CHAPTER 8
THE BATTLE FOR PUBLIC OPINION

1. "Coal Mining in the US—Market Size 2005–2027." *Industry Market Research, Reports, and Statistics*. IBISWorld, 01 Mar. 2021. Web. <https://www.ibisworld.com/industry-statistics/market-size/coal-mining-united-states/>.

2. "U.S. Coal Supply, Consumption, and Inventories." *Short-Term Energy Outlook*. EIA, June 2021. Web. <https://www.eia.gov/outlooks/steo/tables/pdf/6tab.pdf>.

3. "Natural Gas Distribution in the US—Market Size 2005–2027." *Industry Market Research, Reports, and Statistics*. IBISWorld, 29 Mar. 2021. Web. <https://www.ibisworld.com/industry-statistics/market-size/natural-gas-distribution-united-states/>.

   "U.S. Natural Gas Consumption by End Use." *Independent Statistics & Analysis*. EIA, 28 May 2021. Web. <https://www.eia.gov/dnav/ng/ng_cons_sum_dcu_nus_a.htm>.

4. Kagan, Julia. "What Is the Percentage Depletion Deduction?" *Investopedia*. 17 Jan. 2021. Web. <https://www.investopedia.com/terms/p/percentage-depletion.asp>.

5. Redman, Janet, Kelly Trout, Ken Bossong, and Alex Doukas. *Dirty Energy Dominance: Dependent on Denial—How the U.S. Fossil Fuel Industry Depends on Subsidies and Climate Denial.* Rep. Oil Change International, Oct. 2017. Web. <http://priceofoil.org/content/uploads/2017/10/OCI_US-Fossil-Fuel-Subs-2015-16_Final_Oct2017.pdf>.

6. Holland, Stephen, Erin Mansur, Nicholas Muller, and Andrew Yates. *Decompositions and Policy Consequences of an Extraordinary Decline in Air Pollution from Electricity Generation.* Working paper no. 25339. National Bureau of Economic Research, Dec. 2018. Web. <https://www.nber.org/papers/w25339>.

7. Nussey, Bill. "How Much Solar Would It Take to Power the U.S.?" *Freeing Energy.* 06 July 2018. Web. <https://www.freeingenergy.com/a100>.

8. *User Clip: Elon Musk at the National Governors Association 2017 Summer Meeting.* Perf. Elon Musk. *C-SPAN.* 15 July 2017. Web. <https://www.c-span.org/video/?c4676772%2Fuser-clip-elon-musk-national-governors-association-2017-summer-meeting>.

9. *Farms and Farmland.* Rep. United States Department of Agriculture, Aug. 2019. Web. <https://www.nass.usda.gov/Publications/Highlights/2019/2017Census_Farms_Farmland.pdf>.

10. Leirpoll, Malene Eldegard, Jan Sandstad Næss, Otavio Cavalett, Martin Dorber, Xiangping Hu, and Francesco Cherubini. "Optimal Combination of Bioenergy and Solar Photovoltaic for Renewable Energy Production on Abandoned Cropland." *Renewable Energy* 168 (2021): 45-56. Web. <https://www.sciencedirect.com/science/article/pii/S0960148120319236#appsec1>.

11. *Electricity Information: Overview.* Rep. EIA, July 2020. Web. <https://www.iea.org/reports/electricity-information-overview>.

12. Weaver, John Fitzgerald. "Solar Panels Increase Grasses for Sheep and Cows by 90%." *PV Magazine USA.* 12 Nov. 2018. Web. <https://pv-magazine-usa.com/2018/11/12/solar-panel-increase-sheep-and-cow-grasses-by-90/>.

13. Johnson, Scott. "Crops under Solar Panels Can Be a Win-win." *Ars Technica*. 05 Sept. 2019. Web. <https://arstechnica.com/science/2019/09/crops-under-solar-panels-can-be-a-win-win/>.

14. *The Value of Energy Tax Incentives for Different Types of Energy Resources*. Rep. no. R44852. Congressional Research Service, 19 Mar. 2019. Web. <https://fas.org/sgp/crs/misc/R44852.pdf>.

15. "How Much Ethanol Is in Gasoline, and How Does It Affect Fuel Economy?" *Independent Statistics & Analysis*. EIA, 04 May 2021. Web. <https://www.eia.gov/tools/faqs/faq.php?id=27&t=10>.

16. "Feedgrains Sector at a Glance." *USDA Economic Research Service*. USDA. Web. <https://www.ers.usda.gov/topics/crops/corn-and-other-feedgrains/feedgrains-sector-at-a-glance/>.

17. "Corn Usage by Segment 2020." National Corn Growers Association, 15 Jan. 2021. Web. <http://www.worldofcorn.com/#corn-usage-by-segment>.

18. " An acre of solar is 73x more efficient than corn ethanol for cars." *Freeing Energy*. Web. <https://www.freeingenergy.com/m170/>.

19. Wilton, Allison, and John Newton. "Farm Loan Delinquencies and Bankruptcies Are Rising." American Farm Bureau Federation, 31 July 2019. Web. <https://www.fb.org/market-intel/farm-loan-delinquencies-and-bankruptcies-are-rising>.

20. Bookwalter, Genevieve. "The Next Money Crop for Farmers: Solar Panels." *The Washington Post*. 22 Feb. 2019. Web. <https://www.washingtonpost.com/business/economy/the-next-money-crop-for-farmers-solar-panels/2019/02/22/2cf99e8c-3601-11e9-854a-7a14d7fec96a_story.html>.

21. Sönnichsen, N. "Global Oil Reserves Volume in Billion Barrels 2019." *Statista*. 10 Mar. 2021. Web. <https://www.statista.com/statistics/236657/global-crude-oil-reserves-since-1990/>.

22. Jaskula, Brian. *Lithium*. Rep. USGS, 2010. Web. <https://s3-us-west-2.amazonaws.com/prd-wret/assets/palladium/production/mineral-pubs/lithium/mcs-2010-lithi.pdf>.

23. Jaskula, Brian. *Lithium*. Rep. USGS, 2020. Web. <https://pubs.usgs.gov/periodicals/mcs2020/mcs2020-lithium.pdf>.

24. "Tesla Battery Day." EIA/Tesla, 22 Sept. 2020. Web. <https://tesla-cdn.thron.com/static/VM2Z8A_BD_Keynote_final_PSPS0F.pdf>.

25. Yang, Sixie, Fan Zhang, Huaiping Ding, Ping He, and Haoshen Zhou. "Lithium Metal Extraction from Seawater." *Joule*2.9 (2018): 1648-651. Web. <https://www.sciencedirect.com/science/article/pii/S2542435118302927>.

26. "Abundance of Elements in Earth's Crust." *Wikipedia*. Wikimedia Foundation. Web. <https://en.wikipedia.org/wiki/Abundance_of_elements_in_Earth's_crust>.

27. Fischer, Markus, Michael Woodhouse, Susanne Herritsch, and Jutta Trube. *International Technology Roadmap for Photovoltaic (ITRPV)*. Rep. VDMA Photovoltaic Equipment, Oct. 2020. Web. <https://itrpv.vdma.org/documents/27094228/29066965/Update%30ITRPV%302020/2a8588fd-3ac2-d21d-2f83-b8f96be03e51>.

28. *World Silver Survey 2020*. Rep. The Silver Institute, Apr. 2020. Web. <https://www.silverinstitute.org/wp-content/uploads/2020/04/World-Silver-Survey-2020.pdf>.

29. Flanagan, Daniel. *Copper*. Rep. USGS, 2020. Web. <https://pubs.usgs.gov/periodicals/mcs2020/mcs2020-copper.pdf>.

30. *The Future of Solar Energy*. Rep. MIT Energy Initiative, 2015. Web. <https://energy.mit.edu/wp-content/uploads/2015/05/MITEI-The-Future-of-Solar-Energy.pdf>.

31. "Average Power Plant Operating Expenses for Major U.S. Investor-Owned Electric Utilities, 2009 through 2019." *SAS Output*. EIA. Web. <https://www.eia.gov/electricity/annual/html/epa_08_04.html>.

32. Fischer, Markus, Michael Woodhouse, Susanne Herritsch, and Jutta Trube. *International Technology Roadmap for Photovoltaic (ITRPV)*. Rep. VDMA Photovoltaic Equipment, Oct. 2020. Web. <https://itrpv.vdma.org/documents/27094228/29066965/Update%30ITRPV%302020/2a8588fd-3ac2-d21d-2f83-b8f96be03e51>.

Ciez, Rebecca E., and J. F. Whitacre. "The Cost of Lithium Is Unlikely to Upend the Price of Li-ion Storage Systems." *Journal of Power Sources* 320 (2016): 310-13. 15 July 2016. Web. <https://www.sciencedirect.com/science/article/abs/pii/S0378775316304360>.

33. Goldie-Scot, Logan. "A Behind the Scenes Take on Lithium-ion Battery Prices." *BloombergNEF*. 05 Mar. 2019. Web. <https://about.bnef.com/blog/behind-scenes-take-lithium-ion-battery-prices/>.

34. Azevedo, Marcelo, Nicolo Campagnol, Toralf Hagenbruch, Ken Hoffman, Ajay Lala, and Oliver Ramsbottom. *Lithium and Cobalt – A Tale of Two Commodities*. Rep. McKinsey & Company, June 2018. Web. <https://www.mckinsey.com/~/media/mckinsey/industries/metals%20and%20mining/our%20insights/lithium%20and%20cobalt%20a%20tale%20of%20two%20commodities/lithium-and-cobalt-a-tale-of-two-commodities.ashx>.

35. Shedd, Kim. *Cobalt*. Rep. USGS, 2020. Web. <https://pubs.usgs.gov/periodicals/mcs2020/mcs2020-cobalt.pdf>.

36. "Solar Market Insight Report 2020 Year in Review." *SEIA*. 16 Mar. 2021. Web. <https://www.seia.org/research-resources/solar-market-insight-report-2020-year-review>.

37. "If Solar Panels Are Thrown Away, How Much Landfill Is Required?" *Freeing Energy*. Web. <https://www.freeingenergy.com/m134/>.

38. Forti, Vanessa, Cornelis P. Baldé, Ruediger Kuehr, and Garam Bel. *The Global E-Waste Monitor 2020*. Rep. United Nations University, United National Institute for Training and Research, International Telecommunication Union, and International Solid Waste Association, 2020. Web. <http://ewastemonitor.info/wp-content/uploads/2020/12/GEM_2020_def_dec_2020-1.pdf>.

39. "National Overview: Facts and Figures on Materials, Wastes and Recycling." *EPA*. Mar. 2020. Web. <https://www.epa.gov/facts-and-figures-about-materials-waste-and-recycling/national-overview-facts-and-figures-materials>.

40. Weckend, Stephanie, Andreas Wade, and Garvin Heath. *End-of-Life Management: Solar Photovoltaic Panels*. Rep. International Renewable Energy Agency (IRENA), June 2016. Web. <https://

www.irena.org/-/media/Files/IRENA/Agency/Publication/2016/
IRENA_IEAPVPS_End-of-Life_Solar_PV_Panels_2016.pdf>.

41. Wiser, Ryan, Mark Bolinger, and Joachim Seel. Benchmarking Utility-
Scale PV Operational Expenses and Project Lifetimes: Results from a
Survey of U.S. Solar Industry Professionals. Tech. Lawrence Berkeley
National Laboratory, June 2020. Web. <https://eta-publications.lbl.
gov/sites/default/files/solar_life_and_opex_report.pdf>.

42. Korosec, Kirsten. "Redwood Materials Raises $700M to Expand Its
Battery Recycling OPERATION." *TechCrunch*. 28 July 2021. Web.
<https://techcrunch.com/2021/07/28/redwood-materials-raises-
700m-to-expand-its-battery-recycling-operation/>.

43. *An Ex-Tesla Exec's Plan to Recycle Your Batteries. YouTube*.
Bloomberg, 12 Nov. 2020. Web. <https://www.youtube.com/
watch?v=xtVElI1SoRw>.

44. Zayed, Joseph, and Suzanne Philippe. "Acute Oral and Inhalation
Toxicities in Rats with Cadmium Telluride." *International Journal
of Toxicology* 28.4 (2009): 259-65. Web. <https://journals.sagepub.
com/doi/10.1177/1091581809337630>.

45. "Global Lead Metal Production 2020." *Statista*. 24 Mar.
2021. Web. <https://www.statista.com/statistics/264872/
world-production-of-lead-metal/>.

46. "How Much Lead Is Used in US Solar Panels Each Year?" *Freeing
Energy*. Web. <https://www.freeingenergy.com/m143/>.

47. Houlihan, Jane, and Richard Wiles. *Lead Pollution at Outdoor Firing
Ranges*. Rep. Environmental Working Group, 2001. Web. <https://static.
ewg.org/reports/2001/LeadPollutionAtOutdoorFiringRanges.pdf>.
Miller, Chaz. "Lead-Acid Batteries." *Waste360*. 01 Mar. 2006. Web.
<https://www.waste360.com/mag/waste_leadacid_batteries_3>.

48. Fischer, Markus, Michael Woodhouse, Susanne Herritsch, and
Jutta Trube. *International Technology Roadmap for Photovoltaic
(ITRPV)*. Rep. VDMA Photovoltaic Equipment, Oct. 2020.
Web. <https://itrpv.vdma.org/documents/27094228/29066965/
Update%30ITRPV%302020/2a8588fd-3ac2-d21d-2f83-
b8f96be03e51>.

49. Schlenoff, Daniel C. "The Motor Vehicle, 1917 [Slide Show]." *Scientific American*. 01 Jan. 2017. Web. <https://www.scientificamerican.com/article/the-motor-vehicle-1917-slide-show/>.

50. Lapp, Ralph E., "The Einstein Letter That Started It All; A Message to President Roosevelt 25 Years Ago Launched the Atom Bomb and the Atomic Age." *The New York Times*. 02 Aug. 1964. Web. <https://www.nytimes.com/1964/08/02/archives/the-einstein-letter-that-started-it-all-a-message-to-president.html>.

51. Nussey, Bill. "A Simple and Affordable Plan to Address the Intermittency of Solar and Wind." *Freeing Energy*. 23 July 2019. Web. <https://www.freeingenergy.com/solar-wind-intermittency-batteries-curtail-overbuild/>.

52. Ziegler, Micah S., Joshua M. Mueller, Gonçalo D. Pereira, Juhyun Song, Marco Ferrara, Yet-Ming Chiang, and Jessika E. Trancik. "Storage Requirements and Costs of Shaping Renewable Energy Toward Grid Decarbonization." *Joule* 3.9 (2019): 2134-153. 18 Sept. 2019. Web. <https://www.sciencedirect.com/science/article/pii/S2542435119303009>.

53. "What Is U.S. Electricity Generation by Energy Source?" *Frequently Asked Questions (FAQS)*. EIA. Web. <https://www.eia.gov/tools/faqs/faq.php?id=427&t=3>.

54. "Uranium Marketing Annual Report." *Independent Statistics & Analysis*. EIA, 20 May 2021. Web. <https://www.eia.gov/uranium/marketing/>.; Tables 2 and 3.

55. Harding, Robin. "Fukushima Nuclear Disaster: Did the Evacuation Raise the Death Toll?" *Financial Times*. 10 Mar. 2018. Web. <https://www.ft.com/content/000f864e-22ba-11e8-add1-0e8958b189ea>.

56. *The Future of Nuclear Energy in a Carbon-Constrained World*. Rep. MIT Energy Initiative, 2018. Web. <https://energy.mit.edu/research/future-nuclear-energy-carbon-constrained-world/>.

57. "NuScale Power." *Wikipedia*. Wikimedia Foundation. Web. <https://en.wikipedia.org/wiki/NuScale_Power>.

58. Potter, Ellie. "DOE Approves $1.36B Cost-sharing Award to Build First US NuScale Modular Reactor." *S&P Global Market Intelligence*.

16 Oct. 2020. Web. <https://www.spglobal.com/marketintelligence/en/news-insights/latest-news-headlines/doe-approves-1-36b-cost-sharing-award-to-build-first-us-nuscale-modular-reactor-60766451>.

59. Morgan, M. Granger, Ahmed Abdulla, Michael J. Ford, and Michael Rath. "US Nuclear Power: The Vanishing Low-carbon Wedge." *Proceedings of the National Academy of Sciences* 115.28 (2018): 7184-189. Web. <https://www.pnas.org/content/115/28/7184.short>.

60. "My Customers Would Have Asked For a Faster Horse." *Quote Investigator.* 28 July 2011. Web. <https://quoteinvestigator.com/2011/07/28/ford-faster-horse/>.

61. "Plans For New Reactors Worldwide." World Nuclear Association, May 2021. Web. <https://www.world-nuclear.org/information-library/current-and-future-generation/plans-for-new-reactors-worldwide.aspx>.

62. Xu, Muyu, and David Stanway. "China Doubles New Renewable Capacity in 2020; Still Builds Thermal Plants." *Reuters.* 21 Jan. 2021. Web. <https://www.reuters.com/article/us-china-energy-climatechange-idUSKBN29Q0JT>.

63. "U.S. Nuclear Industry." *Independent Statistics & Analysis.* EIA, 06 Apr. 2021. Web. <https://www.eia.gov/energyexplained/nuclear/us-nuclear-industry.php>.

CHAPTER 9
UNLOCKING OUR POWER

1. Tyson, Alec, Brian Kennedy, and Cary Funk. "Climate, Energy and Environmental Policy." Pew Research Center, 26 May 2021. Web. <https://www.pewresearch.org/science/2021/05/26/climate-energy-and-environmental-policy/>.

2. "Omnibus Poll." Huffington Post. Web. <http://big.assets.huffingtonpost.com/toplines_popularthings.pdf>.

3. Kennedy, Brian, and Cary Lynne Thigpen. "More U.S. Homeowners Say They Are considering Home Solar Panels." Pew Research Center, 17 Dec. 2019. Web. <https://www.pewresearch.org/fact-tank/2019/12/17/more-u-s-homeowners-say-they-are-considering-home-solar-panels/>.

4.   Graziano, Marcello, and Kenneth Gillingham. "Spatial Patterns of Solar Photovoltaic System Adoption: The Influence of Neighbors and the Built Environment." *Journal of Economic Geography* 15.4 (2015): 815-39. Web. <https://academic.oup.com/joeg/article/15/4/815/2412599>.

5.   Mikhitarian, Sarah. "Homes With Solar Panels Sell for 4.1% More." *Zillow Research*. 16 Apr. 2019. Web. <https://www.zillow.com/research/solar-panels-house-sell-more-23798/>.

6.   "Cutting the Cord." *The Economist*. The Economist, 07 Oct. 1999. Web. <https://www.economist.com/special-report/1999/10/07/cutting-the-cord>.

7.   "McCaw Cellular Communications." *Wikipedia*. Wikimedia Foundation. Web. <https://en.wikipedia.org/wiki/McCaw_Cellular_Communications>.

8.   O'Dea, S. "Number of Mobile Subscriptions Worldwide 1993-2019." *Statista*. Jan. 2020. Web. <https://www.statista.com/statistics/262950/global-mobile-subscriptions-since-1993/>.

9.   Abzug, Bella S. "Extensions of Remarks." *Govinfo*. 27 May 1976. Web. <https://www.govinfo.gov/content/pkg/GPO-CRECB-1976-pt13/pdf/GPO-CRECB-1976-pt13-3-3.pdf>.

10. "List of Countries by Number of Mobile Phones in Use." *Wikipedia*. Wikimedia Foundation. Web. <https://en.wikipedia.org/wiki/List_of_countries_by_number_of_mobile_phones_in_use>.

11. "Mobile Economy Sub-Saharan Africa." GSM Association. Web. <https://www.gsma.com/mobileeconomy/wp-content/uploads/2020/09/GSMA_MobileEconomy2020_SSA_Infographic.pdf>.

12. Willrich, Mason. Modernizing America's Electricity Infrastructure. MIT, 2017. Print.

13. Mezger, Simon. *Customer Centricity: Must-have or a Waste of Energy?* Rep. Accenture, 2017. Web. <https://www.accenture.com/_acnmedia/pdf-65/accenture-customer-centricity-must-have-waste-energy-pov.pdf>.

14. Damiani, Bailey. "The Best Local Energy Advocacy Organizations in 2021." *Freeing Energy*. 29 June 2021. Web. <https://www.

freeingenergy.com/best-solar-rooftop-advocacy-organizations -2021/>.

15. Data provided by Solar United Neighbors as of July 2021.

16. "Our Impact." Solar United Neighbors. Web. <https://www.solarunitedneighbors.org/about-us/our-impact/>.

17. "Hawaii State Profile and Energy Estimates." *Independent Statistics & Analysis*. EIA, 21 Jan. 2021. Web. <https://www.eia.gov/state/?sid=HI>.

18. "Rooftop Solar Installations up 55% despite Pandemic, Boosting Industry and Hawaii Economy." *Hawaiian Electric*. 25 Jan. 2021. Web. <https://www.hawaiianelectric.com/rooftop-solar-installations-up-55-despite-pandemic-boosting-industry-and-hawaii -economy>.

19. "Alabama State Profile and Energy Estimates." *Independent Statistics & Analysis*. EIA, 20 May 2021. Web. <https://www.eia.gov/state/data.php>.

20. Barbose, Galen, Salma Elmallah, and Will Gorman. *Behind-the-Meter Solar+Storage: Market Data and Trends*. Publication no. DE-AC02-05CH11231. Lawrence Berkeley National Laboratory, July 2021. Web. <https://eta-publications.lbl.gov/sites/default/files/btm_solarstorage_trends_final.pdf>.

21. Howland, Ethan. "Microgrid Tariff Approved by Hawaii State Regulators in Pro Microgrid Move." *Microgrid Knowledge*. 28 May 2021. Web. <https://microgridknowledge.com/hawaii-microgrid-tariff/>.

22. "New 'battery Bonus' Program to Offer Oahu Customers Cash Incentive to Add Energy Storage to Rooftop Solar System." *Hawaiian Electric*. 19 July 2021. Web. <https://www.hawaiianelectric.com/new-battery-bonus-program-to-offer-oahu-customers-cash-incentive-to-add-energy-storage-to-rooftop-solar-system>.

23. "Performance Based Regulation (PBR)." State of Hawaii Public Utilities Commission, 18 Apr. 2018. Web. <https://puc.hawaii.gov/energy/pbr/>.

24. *2008 Republican Platform*. Publication. Republican National Convention, 2008. Web. <https://web.archive.org/web/20090225014449if_/http://platform.gop.com/2008Platform.pdf>.

25. *2008 Democratic Party Platform*. Rep. The American Presidency Project, 25 Aug. 2008. Web. <https://www.presidency.ucsb.edu/documents/2008-democratic-party-platform>.

26. Mildenberger, Matto, Peter D. Howe, and Chris Miljanich. "Households with Solar Installations Are Ideologically Diverse and More Politically Active than Their Neighbours." *Nature Energy* 4.12 (2019): 1033-039. Web. <https://www.nature.com/articles/s41560-019-0498-8#citeas>.

27. Oremus, Will. "The Forgotten Startup That Inspired Google's Brilliant Business Model." *Slate Magazine*. Slate, 13 Oct. 2013. Web. <https://slate.com/business/2013/10/googles-big-break-how-bill-gross-goto-com-inspired-the-adwords-business-model.html>.

28. *National Solar Jobs Census 2020*. Rep. The Solar Foundation. Web. <https://www.thesolarfoundation.org/national/>.

29. "PosiGen—Crunchbase Company Profile & Funding." *Crunchbase*. Web. <https://www.crunchbase.com/organization/posigen>.

30. Barbose, Galen, Sydney Forrester, Eric O'Shaughnessy, and Naïm Darghouth. *Residential Solar-Adopter Income and Demographic Trends: 2021 Update*. Rep. Lawrence Berkeley National Laboratory, Apr. 2021. Web. <https://eta-publications.lbl.gov/sites/default/files/solar-adopter_income_trends_final.pdf>.

31. *Solar Market Insight Report 2020 Year in Review*. Rep. SEIA, 16 Mar. 2021. Web. <https://www.seia.org/research-resources/solar-market-insight-report-2020-year-review>.

32. *Global Off-Grid Solar Market Report Semi-Annual Sales and Impact Data*. Rep. GOGLA, June 2020. Web. <https://www.gogla.org/sites/default/files/resource_docs/global_off_grid_solar_market_report_h1_2020.pdf>.

33. Nussey, Bill. "Samir Ibrahim—The story behind an affordable solar-powered irrigation system that is changing the

lives of smallholder farmers in Africa." The Freeing Energy Podcast, 11 May 2021. Web. <https://www.freeingenergy.com/podcast-samir-ibrahim-sunculture-africa-irrigation/>.

34. "Profile—Samir Ibrahim." *Forbes Lists.* Forbes, 2017. Web. <https://www.forbes.com/profile/samir-ibrahim/?sh=6f36e05d1bca>.

35. Sönnichsen, N. "Revenue of the U.S. Electric Power Industry 2019." *Statista.* 05 Nov. 2020. Web. <https://www.statista.com/statistics/190548/revenue-of-the-us-electric-power-industry-since-1970/>.

36. McCullough, David. *The Wright Brothers.* Waterville, ME: Large Print, 2016. Print.

CHAPTER 10
POWERED BY INNOVATORS

1. Dr. Damiani is best known for inventing and commercializing ion-implantation techniques used by many solar cell manufacturers. Also, full disclosure: Dr. Damiani and the author started a company together in 2018.

2. Weintraub, Seth. "Google's Project Sunroof Data: 79% of US Rooftops Analyzed Are Viable for Solar—Is Yours?" *Electrek.* 14 Mar. 2017. Web. <https://electrek.co/2017/03/14/googles-project-sunroof-data-79-of-us-rooftops-analyzed-are-viable-for-solar-is-yours/>.

3. Henner, Nick. *Energy Efficiency Program Financing: Size of the Markets.* Issue brief. American Council for an Energy Efficient Economy, Nov. 2020. Web. <https://www.aceee.org/sites/default/files/pdfs/energy_efficiency_financing_-_the_size_of_the_markets.pdf>.

4. "Electricity Explained." *Independent Statistics & Analysis.* EIA. Web. <https://www.eia.gov/energyexplained/electricity/use-of-electricity.php>.

5. Geothermal Heat Pumps Market Size, Share & Trends Analysis Report By Type (Open Loop, Closed Loop), By Application (Residential, Commercial, Industrial), By Region, And Segment Forecasts, 2020—2027. Rep. Grand View Research, Aug. 2020.

Web. <https://www.grandviewresearch.com/industry-analysis/geothermal-heat-pumps-market>.

6. "Rivers & Streams." *EPA's Web Archive.* Environmental Protection Agency, 13 Mar. 2013. Web. <https://archive.epa.gov/water/archive/web/html/index-17.html>.

7. *Wind for Energy U.S. Market Analysis.* Rep. One Energy. Web. <https://oneenergy.com/oe-labs/market-studies/>.

8. "Secretary Granholm Launches Hydrogen Energy Earthshot to Accelerate Breakthroughs Toward a Net-Zero Economy." *Energy. gov.* DOE, 07 June 2021. Web. <https://www.energy.gov/articles/secretary-granholm-launches-energy-earthshots-initiative-accelerate-breakthroughs-toward>.

9. Alamalhodaei, Aria. "Swedish Company Northvolt Raises $2.75B to Accelerate European Battery Production." *TechCrunch.* 09 June 2021. Web. <https://techcrunch.com/2021/06/09/swedish-company-northvolt-raises-2-75b-to-accelerate-european-battery-production/?guccounter=1>.

10. "Form Energy." *The Engine.* Massachusetts Institute of Technology. Web. <https://www.engine.xyz/founders/formenergy/>.

11. "Form Energy—Crunchbase Company Profile & Funding." *Crunchbase.* Web. <https://www.crunchbase.com/organization/form-energy>.
Spector, Julian. "Stealthy Storage Contender Form Energy Reveals SECRET Formula: Iron and Air." Canary Media, 23 July 2021. Web. <https://www.canarymedia.com/articles/stealthy-storage-contender-form-energy-reveals-secret-formula/>.

12. Gold, Russell. "Startup Claims Breakthrough in Long-duration Batteries." *The Wall Street Journal.* 22 July 2021. Web. <https://www.wsj.com/articles/startup-claims-breakthrough-in-long-duration-batteries-11626946330>.

13. "Soft Costs Are the Largest Part of Small US Solar Installations." *Freeing Energy.* Web. <https://www.freeingenergy.com/g133/>.

14. "Total Number of Installation Firms for Utility, Commercial, and Residential Solar." *Freeing Energy*. Web. <https://www.freeingenergy.com/ m190/>.

15. National Solar Jobs Census 2020." *The Solar Foundation*. 06 May 2021. Web. <https://www.thesolarfoundation.org/national/>.

16. Full disclosure: the author is related to an executive at Wattch.

17. "Annual Solar Repairs and Maintenance Spend to Grow to $9 Billion by 2025." *Wood Mackenzie*. 22 June 2020. Web. <https://www.woodmac.com/press-releases/annual-solar-repairs-and-maintenance-spend-to-grow-to-$9-billion-by-2025/>.

18. Nussey, Bill. "Are microgrids the future of reliable, clean energy?" The Freeing Energy Podcast, 24 Nov. 2020. Web. <https://www.freeingenergy.com/podcast-ryan-goodman-scale-microgrids-future-clean-reliable-energy/>.

19. McDowall, Mike. "How a Simple 'hello' Became the First Message Sent via the Internet." *PBS*. Public Broadcasting Service, 09 Feb. 2015. Web. <https://www.pbs.org/newshour/science/internet-got-started-simple-hello>.

20. Vailshery, Lionel Sujay. "Number of Connected Devices Worldwide 2030." *Statista*. 22 Jan. 2021. Web. <https://www.statista.com/statistics/802690/worldwide-connected-devices-by-access-technology/>.

21. Alsop, Thomas. "U.S. Households with PC/computer at Home 2016." *Statista*. 12 May 2020. Web. <https://www.statista.com/statistics/214641/household-adoption-rate-of-computer-in-the-us-since-1997/>.

22. "2021 Connectivity and Mobile Trends Survey." *Deloitte Insights*. Deloitte. Web. <https://www2.deloitte.com/us/en/insights/industry/telecommunications/connectivity-mobile-trends-survey.html>.

23. Craig, Ted. "Smart Thermostats Set for Rapid Expansion." ACHR News, 02 Nov. 2020. Web. <https://www.achrnews.com/articles/144003-smart-thermostats-set-for-rapid-expansion>.

24. *Smart Home Report 2021—Energy Management.* Rep. Statista, May 2021. Web. <https://www.statista.com/study/36297/smart-home-report-energy-management/>.

25. Vailshery, Lionel Sujay. "US Smart Speaker Installed Base by Brand 2020." *Statista.* 08 Apr. 2021. Web. <https://www.statista.com/statistics/794480/us-amazon-echo-google-home-installed-base/>.

26. "AutoGrid Flex Platform to Help Sunrun Manage Growing Fleet of Residential Batteries for Grid Services." *Yahoo!* 25 June 2020. Web. <https://www.yahoo.com/now/autogrid-flex-platform-help-sunrun-130000918.html?guccounter=1>.

27. *Report on the Economic Well-Being of U.S. Households in 2019.* Rep. Board of Governors of the Federal Reserve System, May 2020. Web. <https://www.federalreserve.gov/publications/2020-economic-well-being-of-us-households-in-2019-employment.htm>.

28. Cramer-Flood, Ethan. "Uber and Lyft Users and Sales Will Decline Dramatically This Year." *Insider Intelligence.* 22 Sept. 2020. Web. <https://www.emarketer.com/content/uber-lyft-sales-will-plummet-before-rebounding-2021>.

29. " Get Airbnb Statisics For Your Market [2021]." *AllTheRooms Analytics.* Apr. 2021. Web. <https://www.alltherooms.com/analytics/airbnb-statistics/>.

30. Sebi, Carine, and Anne-Lorène Vernay. "Community Renewable Energy in France: The State of Development and the Way Forward." *Energy Policy* 147 (2020): 111874. Web. <https://www.sciencedirect.com/science/article/pii/S0301421520305905>.

31. "Energy Communities." *Energy.* European Commission, 14 Dec. 2020. Web. <https://ec.europa.eu/energy/topics/markets-and-consumers/energy-communities_en>.

32. Rollet, Catherine. "Energy Communities Are Now Allowed in France." *PV Magazine International.* 23 Mar. 2020. Web. < https://www.pv-magazine.com/2020/03/23/energy-communities-are-now-allowed-in-france/ >.

33. Sebi, Carine, and Anne-Lorène Vernay. "Community Renewable Energy in France: The State of Development and the Way Forward."

Energy Policy 147 (2020): 111874. Web. <https://www.sciencedirect. com/science/article/pii/S0301421520305905>.

34. *Peer-to-Peer Electricity Trading.* Issue brief. International Renewable Energy Agency, 2020. Web. <https://www.irena.org/-/media/Files/ IRENA/Agency/Publication/2020/Jul/IRENA_Peer-to-peer_trad- ing_2020.pdf?la=en&hash=D3E25A5BBA6FAC15B9C193F64CA 3C8CBFE3F6F41>.

35. Nussey, Bill. "Will peer-to-peer electricity trading unlock local energy and transform the grid?" The Freeing Energy Podcast, 08 June 2021. Web. <https://www.freeingenergy.com/ podcast-jemma-green-power-ledger-p2p-electricity/>.

36. "Vehicle-to-Grid Technology Market Revenue Is Expected to Reach US\$ 17.43 Bn by 2027." *MarketWatch.* 22 Apr. 2021. Web. < https:// www.marketwatch.com/press-release/vehicle-to-grid-technology- market-share-to-record-us-1743-bn-by-2027-2021-06-25>.

37. *The State of Climate Tech 2020: The Next Frontier for Venture Capital.* Rep. PWC. Web. <https://www.pwc.com/gx/en/services/sustain- ability/assets/pwc-the-state-of-climate-tech-2020.pdf>.

38. Gates, Bill. "Energy Innovation—Why We Need It and How to Get It." *VanadiumCorp Resource Inc.* 30 Nov. 2015. Web. <https:// www.vanadiumcorp.com/news/vanadium-library/battery-research/ energy-innovation-why-we-need-it-and-how-to-get-it/>.

39. "Loan Programs Office." *Energy.gov.* Web. <https://www.energy. gov/lpo/loan-programs-office>.

40. "Portfolio Projects." *Energy.gov.* Loan Programs Office. Web. <https:// www.energy.gov/lpo/portfolio-projects>.

41. "FY 2017 Department of Energy Budget Request Fact Sheet." *Energy.gov.* DOE. Web. <https://www.energy.gov/ fy-2017-department-energy-budget-request-fact-sheet>.

42. Galford, Chris. "USDA Invests \$485M into Smart Grid Upgrades, Rural Electric Infrastructure." *Daily Energy Insider.* 10 Apr. 2019. Web. <https://dailyenergyinsider.com/news/18699-usda-invests- 485m-into-smart-grid-upgrades-rural-electric-infrastructure/>.

43. Cunningham, Lynn, and Rachel Eck. *Renewable Energy and Energy Efficiency Incentives: A Summary of Federal Programs.* Rep. Congressional Research Service, 03 Nov. 2020. Web. <https://fas.org/sgp/crs/misc/R40913.pdf>.

44. Nussey, Bill. "The government's surprising role in super-charging solar technology innovation." The Freeing Energy Podcast, 08 Dec. 2020. Web. <https://www.freeingenergy.com/podcast-becca-jones-albertus-government-solar-seto-doe/>.

APPENDIX

1. Nussey, Bill. "Understanding the Basics of Electricity by Thinking of It as Water." *Freeing Energy.* 02 Nov. 2019. Web. <https://www.freeingenergy.com/understanding-the-basics-of-electricity-by-thinking-of-it-as-water/>.

2. "What Is a Kilowatt Hour?" *Freeing Energy.* Web. <https://www.freeingenergy.com/what-is-a-kilowatt-hour/>.

INDEX